First Published 2005

ISBN: 81-7000-441-1

Ess Ess
Rs.675/-

Published by:
Ess Ess Publications
4837/24, Ansari Road, Darya Ganj,
New Delhi-110 002.
Phone: 51563444
Fax: 51563334
E-mail: info@essessreference.com
url: http://www.essessreference.com

Printed in India at:
Printline, New Delhi

Contents

Preface

Digital libraries are the organizations, which, provide the resources, including the specialized staff to select, structure, offer intellectual access to interpret, distribute, preserve the integrity of, and ensure the persistence over time of collections of digital works, so that they are readily and economically available for use by a defined community or a set of communities. Though, the resources that digital libraries require to perform functions, are similar to those, within conventional libraries, yet, they are, in many ways, different in kind. For storage and retrieval, digital libraries are almost exclusively dependent on computer and electronic network systems.

Like other organizations, digital libraries too need evolve criteria for measuring their performance in a flourishing and highly competitive world. The needs and interests of the communities, they serve will ultimately determine the trajectory of development for digital libraries, including the investment, they make in content and technology.

The present volume provides lot of information on almost all aspects of digital libraries, ranging from imaging technology and techniques to digital library standards, and copyright issues. The content of this work must be valuable to librarians, teachers, researchers and students of Library Science.

Author

1

Key Concepts of Digital Library

The term "digital library" is less transparent than one might expect. The words conjure up images of cutting-edge computer and information science research. They are invoked to describe what some assert to be radically new kinds of practices for the management and use of information. And they are used to replace earlier references to "electronic" and "virtual" libraries.

Digital libraries are organizations that provide the resources, including the specialized staff, to select, structure, offer intellectual access to, interpret, distribute, preserve the integrity of, and ensure the persistence over time of collections of digital works so that they are readily and economically available for use by a defined community or set of communities.

This is a full definition by any measure, and a good working definition because it is broad enough to comprehend other uses of the term. Those other definitions focus on one or more of the features included in the US Digital Library Federation (DLF)definition, while ignoring or deemphasizing the rest. For example, the term "digital library" may refer simply to the notion of "collection," without reference to its organization, intellectual accessibility, or service attributes. This is the particular sense that seems to be in play when we hear the World Wide Web described as a digital library. But the words might refer as well to the organization underlying the collection, or, even more specifically, to the computer-based system in which the collection resides. The latter sense is most clearly in use in the National Science Foundation's Digital Library Initiative. Yet again, institutions may be characterized as digital libraries to distinguish them from digital archives when the intent is to call attention to the differences in the nature of their collections.

The DLF's definition of "digital library" does more than simply enumerate features. It serves in addition as the basis for the DLF's perspective on the scope of digital libraries and on the functional requirements for their development. Brief consideration of certain features of the definition will help to explain its significance to the DLF. <P> *"Organizations that provide the resources...."* Digital libraries are organizations that employ and display a variety of resources, especially the intellectual resources embodied in specialized staff, but they need not be organized on the model of conventional libraries (or even within the context of conventional libraries).

Though the resources that digital libraries require serve functions similar to those within conventional libraries, they are, in many ways, different in kind. For example, for storage and retrieval, digital libraries are dependent almost exclusively on computer and electronic-network systems, and systems-engineering skills, rather than the skills of traditional cataloguers and reference librarians, rank high among the essential staff resources.

Far from emulating the organization of conventional libraries, the organization and structure of digital libraries, and the division of labour within them, are open to considerable experimentation. For example, as publishers and professional societies disseminate works electronically, they are testing how far their investments should incorporate the full range of library functions—and when digital libraries license content from publishers and professional societies that manage their own repositories, they are, in effect, "outsourcing" the library storage function and experimenting with distributed repositories. Further, new organizations appear regularly in the form of small, entrepreneurial, cottage-like industries that scholars, laboratories, and others have developed to create, manage, and disseminate bodies of digital information critical to a discipline or set of disciplines. The physics preprint archive at the Los Alamos National Laboratory is one such development that compels reflection on how digital libraries might best be organized.

"Preserve the integrity of and ensure the persistence...." Each of the functions enumerated in the working definition of "digital library"—select, structure, offer intellectual access, interpret, distribute, preserve integrity, and ensure persistence—is subject to the special constraints and requirements of operating in a rapidly evolving electronic and network environment. The continual change in the environment means that the latter two functions—preserve integrity and ensure persistence—are especially difficult to achieve.

But the DLF regards these functions as central to the concept of "digital library" and follows the Task Force on Archiving of Digital Information in identifying them as linked but distinct. The Task Force argued that the integrity of digital objects is measured in terms of content, fixity, reference, provenance, and context. But it argued as well that the preservation of object integrity, though necessary, is not a sufficient condition of persistence. Persistence depends on other factors as well: organizational will, financial means, and the negotiation of legal rights.

"Collections of digital works...." Distinctions among libraries commonly focus on the subject matter that defines the collections (e.g., medical, art, science, music, and such), or on the communities interested in the collected materials (e.g., research, college, public). The DLF is convinced that, as digital libraries mature, the principle defining their collection policies will not be the "digitalness" of the material. Rather, the defining principles will be, as in other libraries, the subject matter of the materials and the patron community interested in them. The key strategic question for digital libraries anticipating such a development will be how to integrate collections of materials in digital form with materials in other forms. Much of the DLF program seeks to address this critical question.

"Readily and economically available...." Like other organizations, digital libraries need to develop criteria for measuring their performance in an evolving and highly competitive environment. At a minimum, they must reflect the functional attributes of a digital library as described above. One essential measure of the quality of service evaluates performance in terms of cost. Although the costs of digital library service are not yet well understood, the DLF appreciates that successful digital libraries have a sure grasp of critical cost factors and work quickly to economize the influence of those factors. A second essential measure of service quality takes account of how willingly and how responsively a digital library makes information available to its patron communities.

"Use by a defined community or set of communities..." Libraries in general, and digital libraries in particular, are service organizations. The needs and interests of the communities they serve will ultimately determine the trajectory of development for digital libraries, including the investment they make in content and technology. Most of the libraries in the DLF are dedicated to supporting higher education and research, and they justify their investment in digital developments (and in the collaborative work of the Digital Library Federation) as a powerful means of realizing the larger institutional goals of the academic communities they serve.

Architecture of the Digital Library

For the past two years, the Computer Science Technical Reports project (CS-TR) has been developing an architecture for a digital library with funding from the Department of Defense's Advanced Research Projects Agency (ARPA). This is a general purpose framework for a digital library in which very large numbers of objects, comprising all types of material, are accessible over national computer networks. This architecture has been the subject of a series of useful discussions from which eight general principles have emerged. These principles form the key issues in the transition to a true digital library from the network services that we have today. The Kahn/Wilensky paper also contains a comprehensive framework for resolving these issues.

General Principles

The technical framework exists within a legal and social framework

Early networked information systems were developed by technical and professional communities, concentrating on their own needs. The emphasis was on making information available to colleagues and the public, without charge. The digital library of the future will exist within a much larger economic, social and legal framework. For example, musical works represent the livelihood of composers and musicians. Their artistic reputations depend on their work not being changed in storage or transmission. They require payment, as do recording studios and concert halls. Such work will only be part of the digital library, if the library supports their interests.

The legal system's task is to codify this rapidly changing economic and social framework. The relevant areas of law include copyright, performance, and other intellectual property, libel and obscenity, communications law, privacy, and international law. The Kahn/Wilensky architecture can not write the law, but it provides a technical design that matches the legal structure that is expected to emerge. The architecture respects the creators and owners of intellectual property. It allows the preservation of rights that can last for more than one hundred years, and recognizes that digital works may include material from many sources, with separate property rights.

Society expects the creators of works to be responsible for their content, and for those who make decisions about content to behave responsibility. However, the digital library will not thrive if legal liability for content is placed upon parties whose only function is storage and transmission. Therefore, the architecture establishes clear boundaries between the areas of responsibility of the various parties.

Understanding of digital library concepts is hampered by terminology

Terminology proves to be a barrier in describing a digital library. Some words have such strong social, professional, legal, or technical connotations that they obstruct discussion between people of varying backgrounds. Simple words mean different things to different people. For example, the words "copy" and "publish" have different meanings to computing professionals, publishers, and lawyers. Common English usage is not the same as professional usage, and the versions of English around the world have subtle variations of meaning.

Certain words cause such misunderstandings that they are best expunged from any precise discussion of the digital library. The list includes "copy", "publish", "document", and "work". Other words have to be used very carefully and their exact meaning made clear whenever they are used. An example is "content". In the Kahn/Wilensky architecture, items in the digital library are called "digital objects". They are stored in "repositories" and identified by "handles". Information stored in a digital object is called "content", which is divided into "data" and information about the data, known as "properties" or "metadata".

The underlying architecture should be separate from the content stored in the library.

A conventional research library stores more than books, and the digital library is the same. Almost every type of information can be represented in digital form, including text, pictures, musical works, computer programs, databases, models and designs, video programs, and compound works combining many types of information. The underlying architecture of the digital library, as described by Kahn and Wilensky, specifies those characteristics that apply to all types of material. For example, every object needs to have a name or identifier; the actions of adding objects to the library or deleting them apply to all material; general purpose methods of security can be provided.

This underlying architecture is a base for extensions that can be tailored for various types of information. The extensions typically include specific formats, protocols, and rights management that are appropriate for the type of material. For example, the extensions for digitized movies will be very different from those for video games; texts are usually described by bibliographic terms, such as author and title, which are of little relevance to a computer program; a protocol designed for interaction with a database is unlikely to be useful in manipulating graphic designs.

Separating general functions from those specific to the type of content has other benefits. It encourages different markets to emerge, and allows a legal framework in which storage, transmission and delivery of digital objects is separate from activities to create and manage the intellectual content.

Names and identifiers are the basic building block for the digital library

Names are a vital building block for the digital library. Names are needed to identify digital objects, to register intellectual property in digital objects, and to record changes of ownership. They are required for citations, for information retrieval, and are used for links between objects.

These names must be unique. This requires a administrative system to decide who can assign them and change the objects that they identify. They must last for very long time periods, which excludes the use of an identifier tied to a specific location, such as the name of a computer. Names must persist even if the organization that named an object no longer exists when the object is used. There need to be computer systems to resolve the name rapidly, by providing the location where an object with a given name is stored.

The Corporation for National Research Initiatives has implemented a handle system which satisfies these requirements. A "handle" is a unique string used to identify digital objects. The handle is independent of the location where the digital object is stored and can remain valid over very long periods of time. A global handle server provides a definitive resource for legal and archival purposes, with a caching server for fast resolution. The computer system checks that new names are indeed unique, and supports standard user interfaces, such as Mosaic. A local handle servers is being added for increased local control.

Digital library objects are more than collections of bits

In the digital library, information is stored as "digital objects". A primitive idea of a digital object is that it is just a set of bits, but this idea is too simple. The content of even the most basic digital object has some structure, and information, such as intellectual property rights, must be associated with the digital object. One property states whether a digital object is mutable, in that it may be altered after being placed in a repository. Another is a digital signature or other method of validating that an object has not been changed. Frequently, it is useful to keep a log of all transactions associated with each digital object.

The digital library object that is used is different from the stored object

In the digital library, what you store is not what you get. The architecture must distinguish carefully between digital objects as they are created by an originator, digital objects stored in a repository, and digital objects as disseminated to a user. The user receives the result of executing a program on the stored object. This may be a simple program, such as a file transfer program, or something very complex.

For example, an image is stored in a library as a set of wavelets. To use it, the stored wavelets are used to generate an image with the characteristics requested. This is transmitted over the network to a user's computer, where it can be further processed or displayed. Some classes of digital objects can be provided it to a user in more than one way. For example, the score of a musical work is held in the library. One form of use is to transmit a representation of the score to the user's computer.

Alternatively, the user could request the repository to execute a synthesizer program, which would perform the score, and transmit the digitally encoded audio over the network.

For some types of object, such as a data base or a video game, the use consists of an interaction between the user and the execution of the program. Legal scholars see an interesting parallel between the computer viewpoint of executing a program to supply a digital object to a user and the legal concept of performance. This may prove to be the correct framework for managing rights in a digital library.

Repositories must look after the information they hold

A repository stores digital objects, both the content and the metadata. A digital object as stored in a repository may be very different from the digital object that is made available to users' computers. Different repositories will have very different internal organizations, but for each digital object every repository will have a properties record, which holds attributes of the object, and a transaction log.

Since digital objects contain valuable intellectual property, the stored form of a digital object within the repository includes information that allows for it to be managed within economic and social frameworks. The repository maintains this information, provides basic reference information, and provides security to ensure that only valid operations are carried out on the digital objects. The internal organization of a repository and the way that digital objects are stored are hidden from the user.

A simple protocol is provided for interactions with the repository. This protocol is called the "repository access protocol." The basic commands in this protocol are those to access a digital object and its metadata, and the service request to disseminate a digital object. In addition there are commands to add and delete digital objects.

Users want intellectual works, not digital objects

Digital objects are the basic building blocks of the digital library, but users of the library usually want to refer to items at a higher level of abstraction. Common English terms, such as "report", "computer program", or "musical work", often refer to many digital objects that can be grouped together. The individual objects may have different formats, minor differences of content, different usage restrictions, and so on, but certain users are willing to consider them as equivalent.

Which digital objects should be grouped together can not be specified in a few dogmatic rules. The decision depends upon the context, the specific objects, their type of content and sometimes the actual content. The underlying architecture has to support two main needs. It must provide methods for grouping digital library objects and must provide means for retrieval.

The Kahn/Wilensky architecture supports these higher level ideas in several ways. One is to have a digital object containing several digital objects. Thus several formats of a text might be assemble into a single digital object. Another approach is to have these variants stored as separate digital objects, each with its own handle. These handles are contained in a digital object, known as a "meta-object", which acts like a catalogue record. It contains a list of the variants with their handles and information about the differences amongst them.

Usability of Digital Libraries

The term "useful" is generally taken to mean "supporting the required functionality"; in the case of digital libraries, the obvious use is making digital documents available to the appropriate user groups at the time they are needed and in appropriate formats. In practice, users of digital libraries include not just the end users of documents but also librarians and others responsible for the development and maintenance of digital collections, whose needs are centred on collection maintenance. Arguably, if digital libraries are not useful, there is little need to worry about their being *usable*; however, the converse also holds: if a library is not *usable* then it cannot be truly *useful.*

At its simplest, the term "usable" means "can be used". Within Human-Computer Interaction (HCI), "usable" can mean a range of things, including:

— How efficiently and effectively users can achieve their goals with a system (for which it may be possible to apply performance measures);

— How easily users can learn to use the system ("learnability");

— How well the system helps the user avoid making errors, or recover from errors;

— How much users enjoy working with the system — the quality of the user experience — or whether they find it frustrating; and

— How well the system fits within the context in which it is used.

We can consider how these different usability criteria apply specifically to the design and deployment of digital libraries.

Usability Criteria for Digital Libraries

As yet there is no consensus on what the key criteria are for evaluating the usability of digital libraries. The challenge is made more complex by the variety of user types for libraries. For example, some users are the end users of information, finding documents that address their immediate information needs within the context of the task currently being performed, while others are expert intermediaries, performing a document search on behalf of someone else. Digital library users have varying levels of expertise, both in information seeking and in the topic on which information is sought. Also, the tasks of collection maintenance have very different requirements to those of document retrieval and management (from an individual's perspective).

Usability performance measures

Within the broader Information Retrieval (IR) community, two particular performance measures have been used to assess the quality of retrieval algorithms: precision and recall. While the quality of retrieved documents is a key criterion for successful use of a digital library, it is not the only important performance indicator. Indeed, it may not be the best, from a user's perspective. Preliminary studies suggest that users may not want to receive *all* the relevant documents on a topic — after all, they probably will not have time to read them all — but users want to receive a manageable number that can be easily distinguished in terms of the documents' content. While results so far are not conclusive, and it is likely that what is considered a "good" results set will vary with the task and with user expertise, it is already clear that technical metrics do not tell the whole story and that other criteria of quality need to be identified.

Other traditional HCI performance measures, such as average time to achieve goal or mean number of keystrokes, are likely to be difficult to apply meaningfully in digital libraries since time is highly dependent on network performance and many tasks depend on quality of fit between users' search terms and the terms used within the library, rather than on number of keystrokes. However, keystroke efficiency can be used meaningfully for some browse-based tasks. Overall, there is potential for identifying further key performance indicators for usability, but the indicators will never tell the whole story about the quality of a digital library from the user's perspective.

Learnability

Surprisingly little work has been done on how users learn to work with digital libraries, and consequently our understanding of how expertise is developed, and of how digital libraries and their interfaces can be better designed to support learning, is weak. A few studies have compared novices' and experts' information seeking, but without discussion of the design implications of the findings. Informally, novice end users have been found to struggle with learning how to use digital libraries. Libraries demand more sophistication of query formulation than web search engines such as Google™ (http://www.google.com); skills acquired in one library environment are often not easily transferred to another, and interfaces are currently changing rapidly, such that learning can become largely obsolete within a few months. Even professional librarians can find it difficult to keep skills up-to-date.

Preliminary results suggest that end users treat the library system as a tool, not as an object of study. Winograd and Flores extend a discussion by Heidegger on the role of a hammer: skilled users of this tool are largely unaware of it while they are working, until something goes wrong (e.g., the nail bends, or they hit their thumb), at which point their focused engagement with the task is replaced by a focus on the tool and their use of it. In practice, users of digital libraries are not yet able to use them without focusing on the library itself. In contrast, expert intermediaries searching for information on behalf of end users have been found to engage directly with the library, exploring the effects of manipulating search terms on the number and kind of results returned, learning about particular libraries with respect to the current query.

There is clearly much further work to be done on how people learn to search effectively for information, and to use digital libraries in that search — and hence on how libraries can be better designed to support learning.

Errors and error recovery

Arguably, errors are less a problem for digital libraries than they are for systems in which the user is changing data (often irretrievably). However, many user difficulties can be attributed to their making mistakes that they do not even recognise as such and, thus, failing to make progress in their interactions. The quality of feedback from system to user is very often insufficient to ensure users are aware of their actions (erroneous or otherwise sub-optimal) or how their performance could be improved. More critically, while successful searches are often completed quickly, failures are usually protracted, as users persist in trying to find information that either is not available or is not easily accessed. While these incidents are not classic "errors", they result in a poor user experience.

User experience

Even within mainstream HCI, work on user experience has only come to the fore very recently, addressing issues of user pleasure, perceived threats to security and privacy, sense of frustration, etc. Landauer discusses user experience in terms of competence — the user's sense of increasing capability in using systems, which means that users can often gain satisfaction from learning to use inherently poorly designed systems — and effectance — the user's sense of satisfaction at having achieved an interesting effect.

In the context of digital libraries, competence is more often experienced as a negative — as a sense of frustration. At the JCDL 2002 workshop, many participants raised this as a key aspect of user experience. At the subsequent ECDL 2002 panel, Wilma Alexander expressed this from the perspective of a professional librarian, drawing her inspiration from Kipling in structuring her observation of the kinds of questions she gets from users:

— What is this site for?

— Why do I have to login?

— When did they move the website?

— Where is the search button?

— How do I get fewer answers?

— Who is responsible for this site?

In terms of effectance, a good user experience does not seem to depend on always finding the perfect document; but it does require finding new, interesting, and possibly surprising material. The importance of serendipity in search should not be underestimated, although it is difficult to design explicitly to support it. Conversely, failure to find relevant material rapidly leads from frustration to disillusionment to giving up.

Context of use

The final aspect of usability we consider here is the fit of a library within the context of use — organisational, physical, and task-oriented. The use of information resources in various contexts — e.g., legal work, journalism, Humanities research — have been studied by the information seeking community. However, relatively little of this work has related findings directly to the design and deployment of digital libraries. Some of these studies have highlighted the fact that information seeking in context often has a pattern. Initially, users are typically unsure of exactly what their requirements are, and they are therefore unable to clearly assess the relevance of documents retrieved.

The information problem is gradually resolved to a point where users can more clearly define search terms and assess document relevance. Various authors, present models on how users select sources, formulate queries, assess results, etc., but not of the broader information task. Digital libraries typically provide poor support for the earliest problem formulation stage, or for re-finding documents seen but not immediately judged relevant during that stage. It seems likely that a better understanding of how information seeking with digital libraries forms part of the larger information task — whether that be preparing a legal brief, diagnosing an illness or writing an article — would result in the design of libraries that are better suited to the broader task context.

In some situations — e.g., the provision of medical information — the introduction of new ways of accessing information can create conflicts within the organisation. The democratisation of information can break down traditional power hierarchies by empowering junior staff while disadvantaging senior staff who have poorer information seeking skills and less time to acquire them. While recognition of these factors may not lead one to consider detailed design changes to mitigate against or accommodate such effects, the need for sensitive deployment and appropriate user training are clearly indicated.

Just as the roles and relationships of end users within an organisation are changed by the introduction of digital resources, so those of librarians are also going through a period of rapid change. As information becomes more readily available to end users from various locations, those users have less need to go to a physical library, risking leaving the traditional librarian "out of the loop". This demands changing working practices to ensure that communication between librarians (as information specialists) and end users (as consumers of information) remains effective. Arguably, the design of digital libraries should make more provision than is currently the case for end users to access local support for library use.

More generally, end users need information at the right time and in the right place. At present — and probably for the foreseeable future — paper still has an important role to play, as does people's ability to share, annotate and organise pertinent documents in ways that support their broader tasks. Effective digital libraries will be those that support users' information tasks. As one librarian at the JCDL workshop observed, use of any particular digital library is discretionary, and "what they need is information, not to use this library".

Design and Deployment Process

One of the major difficulties of digital library development is that disjointed teams of developers are typically responsible for different components: one group may be designing the core digital library system while another is building document collections, a third is providing a local portal through which to access those collections, and a fourth is delivering web browsing systems. What the user experiences is the overall effect of the interactions between these various system components that may be more or less well integrated.

To date, few studies have investigated the usability of publicly available (e.g., by subscription) digital libraries that are accessible over the Web. The majority of usability studies focus on the usability of one particular library, very often using pre-defined tasks (rather than the user's chosen in-context tasks) for the study. There has also been a greater focus on *post hoc* usability evaluation (after a system has been designed) than on user-oriented tools to support design. Most evaluation studies have been empirical. Some apply quantitative techniques — for example, comparing two versions of a system to establish which is better on some measurable criteria.

Others focus more on qualitative techniques, typically identifying user difficulties when working with a new system. Such techniques generally deliver results that are highly pertinent to the design of the particular system being investigated, but that do not readily generalise to the design of other systems. One great challenge articulated by participants at the JCDL workshop was that of finding and incentivising typical users, and giving them meaningful tasks. Others have sought to reduce the need for representative users by developing heuristics and checklists to enable librarians and other key staff to assess usability without empirical evaluation, or as a precursor to such evaluation. For example, some have applied Nielsen's usability heuristics to the design of digital libraries. Taking a more focused approach, Sandusky presents a checklist of criteria that any digital library should satisfy.

Several librarians at the workshop expressed a desire to have a compendium of tools, together with guidance on what evaluation tools to apply when. Rather than just *post hoc* evaluation, insightful though that can be, it is clearly preferable to build a user perspective into the design process. Various research projects are tackling this in different ways.

One approach, exemplified by the work of Bishop and Bruce is to include end users as members of the design team ("participatory design"). In this approach, the user perspective is represented throughout the design process. As well as the clear benefits of such an approach, there are some well-documented disadvantages. such as those particular users gradually being assimilated into the design team in a way that they lose touch with the perspective of the users they initially represented, and users not being particularly design-savvy. Nevertheless, as an approach it represents a great step forward from non-consultative delivery of technical solutions that fail to match user needs.

Another approach, reported by Jones and Sumner, focuses not on how to evaluate a particular system, but on how to evaluate and improve an organisational structure and development process — a much less well defined problem.

Finally, a third approach being investigated by groups including Bolcini and Paolini and Keith *et al.* is the development of tools to support design. These include the use of Claims Analysis and scenarios, adapted from the work of Rosson and Carroll, which involves members of the design team creating scenarios of use and developing a design rationale, accounting for how well the proposed design will support users in the scenarios.

The effectiveness of this approach depends on the design team being open to critical reflection on their own work, and on the quality of the scenarios used, which are ideally based on extensive data gathering from representative users. There is no "silver bullet" that ensures good user-centred design. The challenges of technical complexity are compounded by the complexities of understanding users' tasks and information seeking behaviours.

2

Digital Library Resources

The online catalogue in libraries is mainly a phenomenon of the 1980s. Only a few academic and some special libraries had implemented them earlier. In mid-1983, Hildreth reported that OPACs were in use in more than 100 libraries in the United States. These and subsequent systems have evolved from several different directions and variety of sources. For example, of 34 OPACs described by Matthews in 1982, about half had been developed by libraries; the other half were developed commercially. OPACs developed by libraries generally are in use only at the originating institution, although an increasing number of libraries gave sold or leased their software to other libraries.

Commercially developed systems are already in use in a substantial number of libraries. Some OPACs in use today emerged from online circulation systems. Probably the best example of a long-standing library-developed OPAC with origins in an online circulation system is LCS at the Ohio State University. In the commercial area, all of the largest-and longest-established online circulation system vendors have developed OPAC modules, including such firms as CLSI, Data Phase, Geac and Universal Library Systems. The evolution from circulation system into a system that incorporates a public-access component has been fairly common in the commercial area, but a number of OPACs developed by libraries were designed first and foremost as online catalogues, independent of circulation functions.

The University of California's MELVYL system and the catalogue Access system of the Minnesota State University. System Project for Automated Library Systems (MSUS/PALS) are two examples. Whichever component of a large-scale, in-house automated system is developed—circulation or OPAC—development of the other may follow. Some of the systems developed by libraries, such as VI designed by Virginia Polytechnic Institute, also incorporate function in addition to circulation and public access; acquisition capabilities are the most common, and some provide serials control capabilities.

Locally designed OPACs are found in many types of libraries, although the most widespread development has occurred in the academic library. Some of the better known include systems developed by Ohio State, the University of California, Northwestern University, the Minnesota State University System, Virginia Polytechnic Institute. Dartmouth College, the Claremont Colleges, Washington State University and Pennsylvania State University OPACs have also been developed in medical libraries, such as systems at the National Library of Medicine (NLM), the Washington University School of Medicine in St. Louis, and the Beth Israel Hospital in Philadelphia. OPACs developed by public libraries have not been as numerous, although the Pikes Peak Library District in Colorado and the Dallas Public Library are examples of public libraries that have developed and currently use the own online catalogue systems. In addition to academic, medical, and public libraries, some smaller special libraries, especially in corporations, use locally developed online catalogues.

Online catalogues can be looked at from several perspectives. Many issues pertain to their design and use in libraries, ranging from the number of terminals required to service a given number of users, to the ergonomics of terminal placement, to the types and amount of information to be included in a catalog, as well as its arrangement on a screen display. But in examining any from of library catalog, online or other, the distinction between its implications for the internal operations of the library and its implications for library users must be kept in mind. Even though these two perspectives do act on one another in the creation and use of the final product, the distinction is important.

The internal-operations perspective has to do primarily with the production and maintenance of a catalog; the library-user perspective has to do with the presentation and use of it. The introduction.of offline, computer-produced book and Computer Output Microform (COM) catalogues was a significant step in the evolution of the library catalog, but almost exclusively in terms of production and maintenance. Computer-produced, offline catalogues addressed many of the problems inherent in card catalogue production and maintenance, and many labor-intensive tasks that were fundamental to card catalogues were eliminated with the use of computers and machine-readable records. With offline computerized catalogues, only one unit entry record had to be created for each item; the computer expanded this record into as many individual catalogue entries as necessary and performed all alphabetization and filing tasks to maintain the catalog.

It was also much easier, at least in theory, to change a major catalogue heading; the change could be made once and applied to all effected entries by the computer-quite a difference from manually typing a new heading onto each which a change applied. However, as far ass the library user was concerned, the benefits of computer-generated offline catalogues were far less pronounced. The actual format of the çatalogue certainly changed in some instances, but the basis content, and the basis avenues through which records in the catalogue could be found, rally did not. The principles used for access to entries in computer-generated book and COM catalogues mirrored those of the card catalog, and the information was often identical and sometimes even less in amount. From the users standpoint, computer-produced book catalogues were little more than card catalogues in bound volumes, and COM catalogues little more than card catalogues in a box.

Given such frustrations as dealing with multiple parts of a catalogue issues in editions and supplements, and equipment breakdowns with COM catalogues, more advantages of computer-produced offline catalogues were not noticeable to the user and, at any rate, could often be overshadowed by the logistical difficulties in using them. In contrast, many OPACs, represent a radical departure both from card catalogues and from offline computerized catalogues in their implications for the user. The relationship between user and online catalogue is fundamentally different from that in use of other catalogue forms.

The online catalogue is interactive, not passive, and requires the user to engage in a dialogue quite uncharacteristic of card, book and COM catalogues. Just to look into the catalogue requires new procedures on the part of the user. A search must be physically, not just mentally, initiated by typing onto a keyboard issuing a command through some other means. The system will respond according to the very specific action initiated by the user. In many instances, the user will then have to initiate further action based on the system's response, and so on back and forth until the user arrives at the desired end of the search or otherwise terminates the dialogue.

This interaction and the nature of system responses represent a significant departure for the user from simply thumbing through catalogue cards or pressing a button, turning a knob, or sliding a handle on a microfilm or microfiche reader. The fact that a OPAC is intended for users of library materials in perhaps the fundamental distinction between this application of automation and those which serve strictly internal processing functions.

Such matters as simplicity of operation or sophistication of retrieval capabilities should be important considerations in the design and selection of any automated system regardless of application, but self-evidence in operation may be less important in a technical processing system in which the library is able to exercise a certain degree of control over the selection, training, and monitoring of system users. The design of an online catalogue must take into account a quite different set of circumstances. Considerable diversity exists among users in their level of knowledge about library catalogues and in the extent to which they are familiar with the use of computerized systems.

The library will usually have little or opportunity to balance its diversity though a planned program of controlled instruction and monitoring. Most people will not use the OPAC as part of everyday routine, and hence their knowledge and skills will not be constantly reinforced. An example of the well-placed emphasis in discussions of OPACs on users and their interaction with the catalogue is the series of studies sponsored by the Council on Library Resources (CLR) during the early 1980s, aimed at gaining a better understanding of what users want and need in online catalogues, their likes and dislikes, and the dynamics of their interaction with catalogues.

Despite the vast array of differences among existing OPACs and lack of specific guidelines for development, at the base of these differences is a common goal: to design online catalogues that will allow the user maximum flexibility and yet will be easy to use. Investigations like those sponsored by CLR should help to establish a stronger foundation for realizing such a goal.

Phrases such as "user friendly" have become almost meaningless because of their sometimes indiscriminate use to describe systems or features that commonsense observation show are 'anything but.' Yet the overuse, and ever misuse, of well-worn phrases does not negate the relevance of the concepts they represent. The degree to which OPACs are willingly accepted by library users will indeed depend in large part on just how user friendly these systems are.

Technical processing systems can be evaluated largely in terms of functional capabilities, or what a system can do. But the circumstances in which OPACs are still used are too different for such a simple evaluation. Equally and perhaps more important is the question of how easily system functions can be learned, understood and applied to the fullest extent by a diverse group of users. For any given characteristics in their operation, OPACs are still simply too new, and their use still too little understood, to be able to state categorically that one approach is somehow "better" than the other. Fundamental to all of the features described is the most basic improvement that OPACs have made on other forms of the library catalog: the expansion of avenues of access to catalogue records. Some of the techniques used to accomplish this are reviewed in the next section.

Access to Online Catalogue

The usefulness of different forms of library catalogues can be compared in a variety of ways. One of the most important criteria is simply the extent to which a catalogue fulfills its basic purposes. Although the purposes of the library catalogue are not case in stone and depend on one's perspective, one set of standards by which library catalogue functions have been measured throughout the twentieth century is the principles of Charles Ammi Cutter as noted in his Rules for a Dictionary

Catalogue. Cutter enumerated three major "objects" of a library catalogue.

— A catalogue should allow someone to find book if its author, title, or topic is known.

— A catalogue should fulfill a gathering function, showing what a library owns by a given author, on a given subject or in a given type of literature.

— A catalogue should provide assistance in the change of a book, either as to its edition or its literary or topical character.

A fourth commonly recognized catalogue function, although not explicitly noted by Cutter, is collection, having to do with the arrangement and relationship of headings within the catalog. Underlying most of these functions is the fundamental matter of access top bibliographic records in the catalog: the points of entry through which a record for an item may be found. Cutter's first object—the finding function—is the essential basis for access, nut the gathering and collocation functions also have to do with accessibility to records.

The function least directly related to access is assistance in the choice of material, which has more to do with content of a bibliographic record than access to it. Whether Cutter's objects still provide the most viable criteria for defining the purposes of a library catalogue in an online era is debatable, and some have questioned the degree of completeness to which these objects have ever actually been realized in libraries. Cutter's principles stem from a time quite different from the 1980s. New factors—including but not limited to those of a technological nature—should not be disregarded in defining the purposes of a library catalog.

Nevertheless, Cutter's principles serve as a point of departure in many decisions of online catalogues, and concerning access to bibliographic records they still provide a convenient framework for comparing online catalogues with one another and with other forms of the library catalog. Access to bibliographic records in library catalogues has traditionally been marked by a high degree of rigidity, requiring considerable knowledge and exactitude on the part of the user.

Generally, access is gained through four types of bibliographic elements:

— author

— title

— series and

— subject.

Multiple author and subject headings are assigned to an item as rules or judgment dictates, but the number is usually kept to a minimum. The terminology selected for headings is also tightly controlled, with the library usually trying to establish a one-to-one correspondence between headings and the persons, corporate bodies, or concepts they represent. These practices may be understandable from a practical point of view, but they do impose certain limits on accessibility to records. The user can indeed locate an item in the collection for which the author, title, or subject is known, but only under some rather inflexible conditions.

To find a record under the title, the user must know the full and exact wording of the title—not some or most of the words, but all of the words, and in their correct order. For access by author, because of the pre-AACR practice of establishing full and singularly valid forms of names, it has not always been enough to know the from of an author's name as it appears on publications or by which the person is popularly known. It has been necessary for the user somehow to know the exact from of a person's or corporation's name as it has been established by the library, often on the basis of considerable research. Where there are discrepancies in name form, it has been up to the user to figure out the form selected by the library.

Access by subject is even more problematic. Singularly valid forms of headings are selected from among all possible equivalents, but because of the greater conceptual complexity involved, there is even greater likelihood in subject searching that discrepancies will arise between the terminology selected by the library and the vocabulary chosen by the user. The problem is further compounded by the general conservatism in libraries toward changing the vocabulary used in headings and by complicated subject heading formation and filling sequences. Such practices, which result in a lack of flexibility, are all but dictated by necessity in card catalogues; however, they were also carried over generally intact to offline versions of computerized catalogues as well. Given the capabilities of computers and the elimination of the need for manual filing, there is no strictly technological reason for some of these practices to have been retained so uniformly.

Additional access points, such as by keywords appearing anywhere in titles or assigned subject headings, could have been provided with little or no appreciable increase in human labor. Similarly, duplicate listings under variant forms of names could have been supplied in the catalogue. These expansions would have required greater sophistication in indexing and would have resulted in a far greater number of entries, and thus larger catalogues, but whether for economic reasons, ties to tradition, or something else, expansions of this type were never characteristic of offline computerized catalogues.

Only with the introduction of online catalogues have such expansions taken firm hold, and they are changing the very concept of access to records in the library catalogue. Some OPACs adhere quite closely to other forms of the catalogue in the ways in which access to records may be gained, requiring just as much exactitude and exhibiting just as little flexibility. These are essentially electronic card catalogues. Other OPACs significantly b.oaden, and even fundamentally change, the traditional concept of access. An over few of several techniques by which this is accomplished is presented in the following pages.

KEYWORDS SEARCHING

A number of OPACs also allow keyword searching on author fields and on fields containing assigned subject headings, and some permit retrieval on the basis of terms appearing within a variety of other bibliographic fields. Keyword searching has certain advantages over the exact order full-phrase searching required in other forms of the library catalogue. In corporate author searches, the keyword capability means that the user does not have to know the exact hierarchy and order in which a corporate author name appears in the records. For title searches, the user does not have to know the full and exact wording of the title to be able to find it in the catalogue.

A key work search capability expands, and actually alters, the traditional concept of locating an item for which, in Cutter's terms, "the author or title is known." In searches for single title or single names, although the entry of selected keyword is certainly less precise than a full title or authors, name and can rest in retrieving extraneous records, among those retrieved will be the desired items. Keyword capabilities are particularly useful in subject searching. When a catalogue permits retrieval of terms appearing anywhere within assigned subject headings, the user does not have to know the full and precise working and filing sequences of assigned subject headings and subdivisions as they appear in the records themselves.

Many OPACs allow a broadened view of the subject elements in a bibliographic record; for instance, by allowing keyword searching on title fields. Given the usually high level of correlation between the subject content of works and the terms appearing in their titles, a keyword title search composed of one or two terms can be a viable from of subject searching in online catalogues. Some OPACs, in fact, automatically retrieve from title fields when subject searches are initiated.

When the user carries out a subject search, records are retrieved for items in which the search terms appear either in assigned subject headings or in the title field of a record. Whether or not this combination is provided automatically by the system, the ability to search assigned subject headings by keyword, supplemented with term searching on other fields of the bibliographic record, makes subject access generally far more expensive in and OPAC than in other forms of the library catalogue. Some OPACs allow a use to truncate, or abbreviate, search terms. For example, many online catalogues allow retrieval of records with an author's surname and first initial.

Even though the full forename may be stored on the bibliographic records, the use of an initial is sufficient for retrieval. In addition to truncation or even omission of forenames and middle names, some OPACs permit retrieval on truncated forms of surnames. Others allow truncation of keywords in titles and assigned subject headings. Entry of the term "prehist" in a title search, for example, will retrieve records containing the terms "prehist" in a title search, for example, will retrieve records containing the terms "prehistoric," "prehistory," or any other derivative in the title. This facility often applies to phrase as well, especially in the case of searching on assigned subject heading fields. Entry of the phrase "Moving Pictures," for example, will retrieve records assigned subject headings "Moving Pictures, Talking," or "Moving Picture—History," and so on.

In a search on assigned subject headings, truncation on most OPACs requires an explicit expression by the user. A special character must be keyed to indicated that a term or phrase is being entered in truncated rather than in complete form. In title searches, some OPACs require the user to supply a special truncation designator; others provide an implicit form of truncation whereby the system automatically truncates a search term or phrase for retrieval according to an internally prescribed number of characters or words, regardless of how many the user has entered. By either means, the net effect of truncation is to broaden the set of records retrieved over what a possible in a search requiring complete and exact terms or phrases.

Most OPACs allow the user to combine two or more search terms through the use of Boolean operators, such as "and," "or," and "not." In most OPACs, a Boolean "and" is assumed when two or more terms are entered in a search statement, especially if the search is conducted of a single field or type of information such as author, title, or subject. Only records containing all specified search terms will be returned. Some OPACs allow the combination of terms from different fields or serving different functions, such as an author's surname and a title word, in which case the user is generally required to enter a economic code preceding each term indicating the field or function to which it applies and then linking the terms with an explicitly stated "and." OPACs with fuller Boolean capabilities allow the user to construct search statements with other operators as well as the most common being "or" and "not."

Depending on which operator is used, the result of Boolean search capabilities is to allow a user to other broaden or narrow the set of records retrieved. Capabilities such as keyword searching expand access to records through traditional avenues like author, title, and subject. Some OPACs index other fields of bibliographic description as well. In a few OPACs, virtually every tern in every field of a bibliographic record are replaced by an all-encompassing concept of access by "bibliographic description" the more specific elements of author, title, and subject headings become simply qualifiers for refining the scope of a search.

OPACs generally have not taken quite this broad an approach, although some have indexed certain additional elements,. such as names of publishers, for retrieval on tern searches that act upon several different fields simultaneously. A number of OPACs offer a more restricted idea of nontraditional access points in the from of bibliographic qualifiers that the user can employ to limit the results of a search. The two most common are language and data of publication. By including a language code or publication data or range of dates in a specified manner as part of a search statement, a user can block the retrieval of records for items that would otherwise qualify but are considered too old or too new to be useful or are in languages that the user does not read.

Other bibliographic elements could be incorporated into OPACs for purposes of qualifying search statements. The possibilities depend largely on the types of information discretely identified and consistently entered into records in the bibliographic database. For libraries with records in the MARC format, one likely source of qualifiers is from the elements constituting the fixed fields. Based on these elements, an OPAC could provide the ability to limit retrieval according to government publication, conference publication, fiction, juvenile literature, and a host of other categories.

Few existing OPACs, however, currently offer a range of qualifiers beyond the elements of language ad data of publication. In a somewhat different but not unrelated category, OPACs have attempted to address traditional limitations of access in library catalogues in the treatment of alternative forms of name headings and subject terminology. The most frequent method used by card and offline catalogues has been to provide cross-references guiding the user from variants in the form of a name or subject term to the authorized, library-established, heading under which full bibliographic entries are listed. As noted earlier, this technique is not altogether service-effective, since cross-reference cards can be "buried" within long lists of entries or complex headings.

Online catalogues generally fall into two categories in attempting to deal with likely alternatives in the terminology through which users might seek access. One technique essentially mirrors the approach commonly used in card catalogues, namely, by cross-references guiding the user to the authorized terminology. When an online cross-reference is displayed in response to a search, the user must then key in or otherwise instruct the system to search under the authorised term or phrase. Although this approach is similar to that used in most other forms of the library catalog, the presentation is usually more distinctive because the cross-reference message will be displayed prominently on a single screen instead of being buried among other information. Other OPACs have taken a different approach with respect to cross-references by essentially dispensing with them from the user's point of view, using instead a recognition of equivalent terms as valid for retrieval.

If the library has assigned "Women" and "Females" as equivalent subject terms in an online authority file, for example, the user will retrieve the same full listing of records in a search on assigned subject headings regardless of which of these two terms is entered in the search statement. This approach still requires the creation and use of an authority file, but the linking of terms is strictly internal and visible to the user. Other features, such as the use of word proximity in term searching, have been incorporated into some OPACs father expand the accessibility to records in the catalog, but those described here are the most common, and they significantly broaden the ways in which library users can approach catalog.

Now of these features is new to online information retrieval systems used in the library environment in general; they have been standard in such systems as DIALOG and BRS for many years. What is new is their incorporation into the public catalogue librarians have long assumed that systems such as DIALOG and BRS require special expertise and training for effective use. What is most challenging about OPACs is that at least some of these same sophisticated features are intended for people without formal training and with only limited need to use them. Taking these factors into account is perhaps the most crucial element in the successful design of online public-access catalogues and their implementation in the library environment.

Because they are interactive and can offer far broader approaches to access, OPACs are a fundamentally different from of catalogue than their predecessors. The overall difference can result in greater value for the library user, but not without costs in human terms. Interaction with computerized systems and the use of such techniques as keyword searching, truncation, Boolean combinations, and so on, involve skills that many users do not have, especially within the context of application in a library catalog.

For many, lack of familiarity with computers and retrieval techniques extends well beyond use in an Online library catalog. There are still many library users for whom interaction with a compute for any purpose remains an alien experience. This situation is changing rapidly as more and more of the population become educated in computer use, but for some time to come, unfamiliarity is a condition that must be taken into account in the design of online catalogues.

Interaction of User/System/Catalog

Using an OPAC requires the interaction of three primary participants: a library user, a computerized system, and a library catalog. With an OPAC as both a system and a catalog, the rules governing behavior can be fairly mechanically and consistently structured, even if at times they are complex. As an online computerized system, an OPAC incorporates certain rules that dictate how a user must enter commands and how the system will react to them.

In the library catalogue function, certain aspects of interaction will be governed primarily by rules and conventions adopted by the library regarding descriptive cataloging, summation of subject content, and so on. Rather than distinguishing between on OPAC as a library catalogue and as the mechanical means of conveyance, some discussions blend the two functions together. However, it is important to recognize such a distinction on at least a conceptual level, for each purpose has characteristics that can independently contribute to success or failure in using an OPAC.

Some patrons may be well versed in the use of computerized systems and their retrieval techniques, but may be almost totally unfamiliar with the library's cataloguing and other bibliographic practices that do, in fact, influence the degree of success in using the system. Other patrons may have a fairly comprehensive understanding of the structure and organization of library catalogues, even in online form, but may be unskilled in the mechanics of using computerized systems and unfamiliar with retrieval techniques. Any advantages gained by introducing clarity, case of use, and greater flexibility at one of these levels can be offset by undue complexity or lack of clarity at the other.

By taking advantage of the power of the computer, an OPAC can be more flexible than conventional catalogue forms, but it must be so in a way that allows these new methods to be easily learned and understood. Although the rules that govern the behavior of an OPAC as a system and as a catalogue can be fairly mechanical and consistent, the rules that govern the behavior of the library user are not. First and always, "the library user" is an elusive entity. Human factors that influence online catalogue use, such as levels of knowledge, skill, and motivation, vary widely from one user to another, and even within the same user at different times.

Complicating matters is the user's perception of how an OPAC works. It is one thing to say that an online system performs a certain way or that certain rules govern the structure and organization of a library catalog, but the more important consideration is the user's own perception about how the system works or how the catalogue is structured and organised. The design features of an OPAC can either facilitate or inhibit interaction between user and system. As noted earlier libraries are not always able to provide the desired level of guidance and assistance when making as significant a change as introducing a sophisticated online public-access catalog.

Therefore, an OPAC must be designed to minimize the need for offline guidance, taking on some of the instructional responsibilities and burden of assistance itself. An OPAC must be easy to use and its features easy to learn, and it must be a free as possible from retype language, confusing conventions, and complex rules that are meaningful only to the system. The term "user friendly" may be ambiguous and trite, but the concepts it represents are perhaps the most crucial criteria for evaluating systems designed to handle the interaction between a computer, a library catalogue, and a diverse group of library users.

There are certain similarities from one online catalogue to another in terms of the sequence of stages in which dialogue occurs. For most OPACs that are not entirely menu-driven, dialogue between a user and the online catalogue generally conforms to the following:

— USER: Enters a command to begin session

— OPCA: Acknowledges user's command.

— USER: Enters a search statement.

— OPAC: Provides "first-level retrieval" of records conforming to the user's search statement.

— USER: Manipulates and refines results of first-level retrieval in order eventually display record(s) desired.

— OPCA: Responds accordingly to manipulation and refinement until the user terminates the session or enters a new search statement.

— USER: Enters command to terminate session.

— OPAC: Acknowledged user's command.

OPACs vary in the extent to which each stage in this general sequence is explicitly necessary, especially in the earliest and latest steps—the begin-session, file-selection, and terminate-session procedures. Many OPACs do not require the user to instruct the system that a new search session is about to begin or to explicitly select the catalogue file. Instead, the OPAC is constantly set for a bibliographic search, enabling the user to begin searching immediately on arrival at the terminal.

Similarly, many OPACs do not require the user to explicitly information system when a session is completed. The user terminates the session simply by walking away from the terminal, and the display remains on the screen until a new search statement is entered. Even when specific begin-session or end-session procedures are required or available, their execution is normally simple and straight forward. The real substance of interaction occurs between the time a user enters as search statement and the time at which he or she feels the search is completed.

The end of a search may be the display of a single bibliographic record for an item, a list of abbreviated bibliographic records containing one or more of the desired items, or a response or display that leads the user to conclude that the item or items are not in the catalog. The dialogue from entry of a search statement through final resolution may require any number of user-initiated actions and system responses. The more specific manner in which various OPACs handle the stages outlined above is examined in the next chapter. But in terms of the general nature of the user interface, it is important here to moment on two of the stages in particular. The first involves entry of a valid search statement and the initial system retrieval of appropriate records.

In some OPACs, this stage consists of a single step by the user and a single response by the system. In other, there are two or more interactive steps involved. For example, the use may first be required to indicate the type of search being conducted—author, title, subject, whatever—to which the system responds by acknowledging the type and issuing instructions to enter search terms. The user then keys in the appropriate term or terms, and the system responds by providing first-level retrieval of all records in the database matching the two-step search statement. Some OPACs, depending on the complexity of the search statement, require even more that two interactive steps before the statement is complete and records are retrieved.

The second major stage of interaction involves the manipulation and refinement of the search results provided by the system at the first level of retrieval. If the search is satisfied at the first level, this will be unnecessary. In many cases, however, the search is not satisfied at first stage, and final resolution may require several steps in manipulating and refining the initial results. For example, a user may seek full bibliographic detail about a specific work, but can remember only one or two words in the title. If the words are very general, for example, "History" or "Education," a keyword title search will retrieve a large number of records at the first level.

To identify the specific work and arrive at a full bibliographic display, the user may have to wade through a rather lengthy discourse worth the system. These two crucial stages of interaction—entry of a search statement and manipulation and refinement of first-level results—are in fact, part of a continuum in interaction worth an online catalog. The distinction between them is made partly on the basis of convenience for purpose of discussion, but also partly on the basis of the crucial role that first-level retrieval plays in establishing the content for subsequent interaction. There are few areas in OPAC design that exhibit as much diversity as does presentation of the initial system response to a search statement.

First-level retrieval in some OPACs involves simply an indication of the number of records retrieved, even if there is only. The user must then enter another command to actually see any bibliographic information at all. Another approach indicates the number of records retrieved, but immediately displays bibliographic information only one record is retrieved. Other OPACs display at least some bibliographic detail at the first level, even if a number of records are retrieved. The way in which this is done, however, can very considerably. For searches where very large numbers of records are retrieved, some OPACs group individual records at a general level if necessary so that everything can be summarized on a single screen display.

In others, the first level consists of brief bibliographic information about each item, regardless of the number retrieved or the number of screens required for brief display, so long as a system limit is not exceeded. In some OPACs, the system automatically scrolls through the entire list of enters until the user initiates some action to freeze the display. In other, a single "page" of entries is fixed and the user must then enter a command in order to see subsequent pages of information, one at a time. A variable in this approach is the criterion used for arranging individual entitles in the list. The basis for arrangement of records in such a listing can vary from one OPAC to another.

Entries are displayed in so online catalogues alphabetically according to author, title, or subject heading. The primary basis in other is reverse chronological order by datas of publication. In others, the basis is by some principle strictly internal to the system or something that may not be immediately apparent to the user, such as sequentially by system-supplied bibliographic record control number. These examples underscore the diversity among OPACs at this crucial stage of interaction. Whatever method is employed, the first level of retrieval is the starting point from which interaction between user and online catalogue will progress.

For any of the stage of dialogue outlined previously, there are specific issues that have a direct bearing on the degree to which an OPAC can clarify and assist interaction between a user and the catalog. Some of these are examined in the next chapter. But underlying the specific issues are more general concerns that are pertinent to the user/system/catalog interactive process regardless of the stage at which interaction is occurring. Among the more important of these general concerns are the explicit features designed to assist the user in conducting dialogue with the system, the language and related conventions employed in OPACs, the nature of the progression of dialogue, and the extent to which a system accommodates different levels of users.

User-assistance features

The length and complexity of dialogue can vary from one OPAC to another and from one search session to another on the same system, but a requirement of any OPAC is that the dialogue must be carried out according to a specific set of rules and within the range of options designed into the system itself. At least at this time, conversation with any OPAC is not entirely free-form.

An OPAC will understand only certain instruction expressed in a prescribed way, and often only at certain points within the course of the dialogue. A command that is valid at one step in the dialogue may not be valid at another. Thus, the range of commands available, the manner in which they must be expressed, and the timing with which they may be used are all strictly defined by the system, For meaningful dialogue to occur, the burden of adjusting to all those limits is on the user.

It is the user who must learn, understand, and confirm to the syntax and semantics of the system, not the other way around. Despite occasional protests to the contrary by some OPAC designers, the rules of operation and conventions of dialogue are not self-evident in most OPACs. A user does not simply intuit the available commands or how and when they are to be entered. They must either be learned obviously presented to the user on a step-by-step basis during each search session. In response to these needs, OPAC designers have tried to build in various online mechanisms to assist a use in carrying out his or her portion of the dialogue. Some of these address short-term and very specific needs, informing the user at any particular moment in a dialogue about the options available at that point and how they mat be exercised. Others are more general or long term.

The form in which, and the extent to which, such features are provided very tremendously among OPACs, but they can be grouped into six general catalogues :

— General explanatory messages.

— Search history: messages indicating what has been done at a previous step or steps during a search.

— Options and prompts: messages indicating what options are available for the user's next action and how they may be exercised.

— Error detection: messages displayed in response to an invalid command entered by the user.

— HELP screens.

— Online tutorials.

User assistance features do not always fall distinctly into one and only one of these categories. For example, a message explaining a user error might also outline the options available for the user's next action; a message in response to a valid command that outlines the options available at that point might also explain what was just done. More often than not, however, these different types of messages and the user-assistance features they signify are independently identifiable, even when two or more are represented on a single screen display.

Most OPACs provide messages outlining the general limits of using the system. Frequently, these messages are concentrated on an introductory screen displayed at the beginning of a search session. They commonly carry the following types of information:

— Generally library data: name of library, hours the system is available and so on.

— Files available on he system (library catalog, community calendar, and the like).

— Scope of coverage of the bibliographic database.

— List of commands or features that may be used during a search session.

— Options available for searching (author, title, subject, etc.).

— How to enter a search statement.

— How to retrieve detailed online HELP screens.

— Where to obtain offline assistance.

No existing OPAC can provide all of these types of information on a single screen display, and some types are less commonly included than others. A list of files available on the system, for example, is necessary only on multipurpose system in which the user must select the library catalogue file as opposed to some other. Although general explanatory messages are usually emphasised on introductory displays or outline in fuller detail on separate HELP screens, some may be included on displays at stages of the dialogue as well. This is most often in the form of message reminding the user of the major options available for searching, the procedures required to enter a new search statement, or the command required to retrieve a HELP screen either general in content or specific to the step that has been reached in the dialogue. An example of this type is shown in.

Search history: Some OPACs can keep track and inform the user of dialogue that has occurred up through any given point in a search. Few, however, have this capabiiity beyond repeating, as part of a system response, the single most recent command entered by the user; for an example, A number of OPACs, in fact, do not even go so far as to repeat the most recent command; instead, they simply display a response without repeating the command or terms that led to that response. Not surprisingly this can lead to confusion, especially if the user has made an error in entering a search term.

If the search term was not repeated as part of the system response, the user would probably surmise either that he or she misspelled the term or that the library did not own any materials by an author whose surname was Commoner. One cannot assume that the user would conclude the former over the latter. Even in OPACs that so repeat the user's previous command as part of the response, there is always the possibility that the user will not read the screen thoroughly enough to realise that a term was misspelled or that a command was entered inappropriately, nut such a feature at least attempts to give clearer explanations of system response to user commands.

Options and prompts: Along with each response to a valid user command, some OPACs display messages that can assist in selecting and executing the next step in the dialogue. In some cases, these messages only inform the user of which options may be exercised at that point. In others, the actual procedures required to exercise each option are also explicitly stated. The types of messages that provide either or both levels of assistance generally fall into one of three major categories.

— Questions prompting the user to answer yes or no.

— Lists of alternative actions that can be taken at that point. Sometimes including how to exercise each.

— Textual or other prompts that indicate where a command is to be entered.

Error detection: When unfamiliarity with computerized systems in general, and online catalogues in particular, is combined with the stringency of rules and conventions required for dialogue with an OPAC, one result is that there is bound to be a certain amount of miscommunication and misunderstanding between the user and the system. The ways in which a user contributes to a breakdown in dialogue vary in degree and kind. So do the ways in which the system responds. In terms of system response, the primary distinction is whether the system can detect that the user has made a mistake.

If not, the request will be processed as though correct; for instance, when a user simply misspells a search term. If the command itself was structured correctly, an OPAC cannot tell tat the search term was misspelled, and it will attempt to retrieve records and display a response on the basis of the misspelled term. For certain types of mistakes, however, the system may be able to recognize that the user has made an error. This usually occurs when a user has misspelled the command name itself (as opposed to just the search term), has tried to use a particular command at a point in the dialogue where it is not allowed, or has entered more search terms than are permitted in a single statement. These and other possibilities have common source: The user has made a request that the system recognizes as contradictory to the basis set of rules and conventions required for dialogue.

The system does not full understand what the user is trying to do. It cannot process the request into normal fashion, but it can detect the fact that miscommunication has occurred. The exact forms of system response under such circumstances vary, but they generally fall somewhere within the following four levels. The system

— recognises that an error has been made and responds in a prescribed manner;

— responds by including an explicit acknowledgment that an error had been detected;

— attempts to explain why it considers the user's request to have been an error;

— suggests what corrective action should be taken.

The relationship amount these levels is usually hierarchical. It would be possible, of course, for an OPAC to respond to an error only by suggesting corrective action, without explicitly stating that an error had been made and without any attempt to explain why the user's action was considered an error, but this is unusual. In practice, a response at one of the four levels almost always includes action or explanation pertinent not only to that level but also to each level preceding. In general, assistance to online users in response to errors is not well developed in OPACs.

A number of OPACs do not interact at even the second level of response. One OPAC, for example, merely displays the "begin-session" screen when an error is detected, regardless of the point to which the search has progressed, and without any explicit statement that an error has been made. Others respond simply by redisplaying whatever was on the screen at the time that the erroneous command was entered, again without explicit statement that an error has been made. Others respond simply by redisplaying whatever was on the screen at the time that the erroneous command was entered, again without explicit acknowledgment of an error or statement explaining why the system responded in that manner.

Most OPACs, however, do respond at least at the second level of error detection, displaying a new screen that includes an explicit statement noting an input error. Some stop at this level, with no further elaboration about why the command was in error or how it could be corrected, thus giving the user only minimal assistance. The language in which the acknowledgment is stated is often cryptic if not downright unsettling for an inexperienced user. Messages such as "Error Type 23" or "You have entered an Illegal Command" are likely to be discouraging two the novice, whose level of confidence in approaching the catalogue may not have been very high in the first place.

More and more, efforts are being made in OPAC design to provide error messages that are meaningful in content and positive in tone, explaining the nature of an error or indicating why the system is unable to process a command, and in some cases even suggesting what corrective action should be taken. You have attempted to change the TYPRUN OPTION, whose value is fixed for your terminal and cannot be changed. The value of the TYPRUN OPTION for the terminals is "$TYPRUN."

— The format requested was unclear. The command should be entered:

— Format: D Set #/Format/Display Option

— Example: D 1/LTTL/ABST

The first example does in a way attempt to explain why the system considers the user's command to be in error, but its content is about as comforting, and as useful, as an "Illegal Command" message. The second example has detected the nature of an error and has even shown the user how to format the command properly, nut the corrective action is still offered in summary OPAC language rather than being explained in easily understandable language. Regardless of the clarity and comprehensiveness with which OPAC massages explain errors detected by the system, there is still the question of miscommunication that is not automatically detectable.

A user may enter a command or terms in a search statement that are perfectly valid from the system's point of view, but that do not really represent the action the user wishes to take. Because a system cannot provide information about, or suggest corrective action to, mistakes that it cannot detect, these sorts of errors are potentially the most damaging to a dialogue. Some techniques, such as repeating the user's command as part of the system response, can provide some help, but they still really solely on the user not the system to recognize the mistake. A rather exceptional feature has been incorporated into the OPAC at the Washington University School of Medicine Library, which attempts to provide at least some level of error detection beyond the fundamental rules and convention required for dialogue.

In the event that no records are retrieved on a search term, the system systematically modifies the user's term by omitting vowels, double consonants, and final's, and then searches the modified term against index terms that have been created similarly. The result is that, under certain conditions, appropriate records can be received even when a search term has been misspelled. This feature, while certainly not foolproof, goes one step further than most OPACs in attempting to incorporate greater tolerance of user errors into the system.

No matter that sophisticated features are included, OPACs will not automatically detect all types of user mistakes, but this problem is certainly not unique to online forms of library catalogues. And, in fact, OPACs have a far greater capacity for recognizing user errors than do any other forms of the library catalogue. Most POACs provided some degree of online assistance in the form of special HELP screen, which contain information describing various aspects of using the system. Their function is not unlike that sometimes provided on general "begin-session" displays or in informative error messages, but these screens are usually displayed in response to a specific, user-initiated request for aid.

The available number of HELP screen ranges from only one or a few on some OPACs, to 30 to 40 on others, to more than 150 on the University of California's MELVYL system. On OPACs that provide comparatively few, the instructional data are usually quite general and related to major stages of a dialogue. On POACs with a greater number of HELP screen, explanation is usually offered on several levels ranging from the very general to the more specific. The more specific levels of detail may provide guidance related to the use of particular commands, access points, or system features such as Boolean searching or truncation, and they sometimes pertain to specific points that have been reached in a dialogue.

The HELP screen repeats the entered subject term and suggests possible reasons why no records were retrieved. If the user concludes that the first reason was the source of the problem, he or she might take the OPAC's advice and enter the command "s" to see a more complete explanation about subject searching for materials in the main library. Of all existing OPACs, the MELVYL system on the University of California campuses has the most extensive array of HELP screens. Many pertain to specific circumstances and are most relevant after certain prints of progress in a dialogue. When requested in reaction to a system-supplied error message, the data on some HELP screens are somewhat diagnostic.

It is extremely important that, after display of either a user-requested HELP screen or a system-initiated error message, the point of progress in a dialogue is not lost. The user should be able either to return immediately to the point in the dialogue reached before a HELP screen or error message was displayed, or even to proceed from the HELP or error display directly to the next logical step. Although this capability is fairly common to OPACs, in some a request for HELP or a system-detected error will mean that the user has to begin the search all over again. Probably the most elaborate from of online assistance that OPAC can offer is programmed tutorial instruction.

Rather than begin problem-or situation-specific, an online tutorial is more a teaching facility than a "moment-of-need" reminder. Few existing OPACs currently provide a fully developed online instructional package. Although such a feature might be useful, it does presume two conditions. First, it implies that the rules and conventions required for dialogue are too complex, or at least not self-evident enough, for effective communication without formal instruction. Second, its value would ultimately depend on the inclination of users to take advantage of it. Both of those condition run counter to trends detected in most OPAC-use studies.

Users generally prefer to be led through a dialogue on a step-by-step basis, getting point-of-progress assistance only when necessary. They do not want to spend a great deal of time and energy reading about how to use the system and memorizing rules and conventions. Some users might indeed welcome such features and would be glad to use them, but OPAC designers should never view online tutorials as a substitute for clarity and completeness provided as part of routine system response during interactive dialogue. Such a shifting of priories would ultimately detract from the overall suitability of an online catalog.

An online tutorial can be of value, but only if intended as additional to, not in lieu of, other types of online user assistance.

Each of the system features already described can play a crucial role in determining the ease with which a user can carry out a dialogue with an online catalog. But merely incorporating any or all of these features into an OPAC does not ensure their usefulness. The manner in which they are presented in equally important, and one of the most critical elements in this is language; the terminology used to convey message, to prompt commands, to explain options, and so on.

The difference between a feature itself and its mode of expression can be illustrated by an example cited earlier. An error message was displayed informing the user that he or she had attempted to change the TYPRUN OPTION, whose value was fixed and could not altered. The user probably had not the faintest idea of what a TYPRUN OPTION was, much less wanting to change it. From the message displayed, the user may have been able to surmise that the intended command had been entered incorrectly, but that conclusion was not part of the actual language used in the error message. That example may be extreme, but in fact most OPACs so use a great deal of specialized system and other jargon.

Generally the language carries a certain degree of ambivalence in basis approach. There is a commitment to employ "natural-language" vocabulary, both expressed by the OPAC and required of the user, that as closely as possible resembles everyday conversational language. At the same time, there is also a strong and visible reliance on language that is not natural in any common conversational sense. It is, rather, "inside" language, specific either to the OPAC itself or in some cases to computerised systems or library catalogues in general. To understand inside language, the user must be familiar with it or clever enough to figure it out.

In either case, at some point the user must learn new vocabulary that may have little or no application outside the specific context of the OPAC itself. The specialised meanings of natural-language terminology in OPACs apply not only to "system" vocabulary, but also to the vocabulary of librarianship. Phrases such as "main entry," "copy number," "full bibliographic record," "Library of Congress Subject Heading," and many others displayed on OPACs all use fairly common terminology, but the way in which the words are combined probably not obvious to most users. Even such an innocuous term as "author" can cause confusion under some circumstances.

In OPACs with the traditional distinctions of author, title, and subject for searching, for example, a user who wishes to see a list of books about rather than by an author must treat the author's name as a subject term rather than as the as the name of an author. This may be an easy distinction for librarians, but it may not be self-evident to all OPCA users. In judging the appropriateness of terminology for presentation in OPACs, librarians should not only be wary of vocabulary that seems somewhat "computers," but should take into account the effects of using their own jargon as well. The language used in some OPACs draws appreciably on non alphabetic characters, both in the presentation of information and in the command structure used in dialogue.

In searching an online catalogue users may have to treat certain characters differently from the method that is usual in searching other forms of library catalogues. Most online systems, for example, recognize some nonalphabetic characters, particularly numerals, as valid in themselves. Thus, in searching of the title 50 Houses of Charleston, South Carolina, users of some OPACs may have to enter the first word as 50, not fifty. This runs exactly counter to the standard procedure in searching in card catalogues. Punctuation within search terms can be another problem, and some OPACs have strict rules governing its use.

In dealing with words, phrases, or abbreviations such as O'Hara, Al-Hardalo, I/O, or I. W. W. as search terms, some OPACs retrieve records only when the punctuation within the term has been entered in the search statement exactly as it appears in the appropriate field of the machine-stored bibliographic record or records. This places something of a burden on the user; he or she may not know the exact punctuation, and is certainly unlikely to know how it appears in the record until the record is actually retrieved.

In contrast, some OPACs do not allow any punctuation within search terms, even though the punctuation will be displayed on the retrieved records. Still others are flexible, retrieving records whether the user includes or omits the appropriate punctuation within search terms. Some OPACs rely heavily on the use of punctuation to separate major elements of information in search statements and bibliographic displays. Users frequently must separate elements of a search statement—for example, the command name or functional qualified from the actual search terms—by some form of punctuation, most commonly a diagonal slash (/) or equals sign (=).

In OPACs that permit Boolean searching, special provisions—sometimes in the form of punctuation—must be made for the user to differentiate between the function of certain words as Boolean operators and as normal search term, he or she must place quotation marks around the world in the search statement (as "and"). In some OPACs, forms of punctuation are used to separate elements in bibliographic displays. In one OPAC, for example, imprint data are enclosed between a less-than sign (<) and a greater-than sign (>), and subject heading are separated by a diagonal slash followed by a number sign (/ #).

Although this practice tends to clutter the display with nondescriptive characters when used in excess, this use of punctuation does separate desecrate pieces of information and is not really a problem unless the user infers that the characters signify something more than simply demarcation. Most OPACs attempt to use natural, easy-to-understand language in explanatory messages, descriptions of options, bibliographic displays, and so on, but they have not broken entirely free of cryptic codes, computers and technical library jargon.

In fact, it is doubtful that they could, or should even try to, free themselves entirely of specialized language. Terms such as enter, entry, exit, record, index, and so on, will always have different meaning in that context asks no more of catalogue users than might be required when applying the terms in other contexts. Clear and understandable language should be part of the pursuit of user-friendly OPACs, but individuals will always have to learn certain precise meanings to carry on dialogue with an OPAC. This is not to say that OPAC languages should not be refined; on the contrary, there is still a great deal of room for improvement.

The language used in OPACs must be balanced between the need for precision in a specialized context on the one hand, and the case with which it can be learned, understood, and applied on the other. Many of the features already described relate to the clarity with which an OPAC can assist a user in progressing from one point in a dialogue to another. A related but some what broader question concerns the ease and speed with which a user may move around among various points or change the direction of a dialogue.

In general, OPACs impose fairly stringent limitations on the number of directions that a dialogue may take from a given point. At any juncture, the user will usually have two or more immediate options, but these will not necessarily include all the options that a user might see as the next logical step for his or her own purposes. The user will always be able to move the dialogue to a point where that desired option may be exercised, but it may require several separate steps to do so. For instance, an OPAC may contain the holdings of several different libraries. A user may complete an author search in one library's catalog, but find no records under the author's name.

If the user wishes to search the same author in one of the library's catalogues, several steps might be required just to get to the point where the same name can be searched in the other catalogue. The speed and case with which an OPAC allows a user to mover in various directions during a dialogue are dictated by design. In the example just given, greater complexity in progression is caused in part by the fact that the OPAC is multi-institutional, containing the catalogues of several different libraries. In a multi-institutional, containing the catalogues of several different libraries.

In a multi-institutional OPAC in which so many steps are required to execute the same search in different libraries' catalogues, design assumption would have to had been made that a user was much more likely on want to search a single library's catalogue for different items than different libraries' catalogues for the same item. But even in OPACs where dialogue does not have to take into account such factors as switching from one library's catalogue to another's, the progression through which dialogue must occur is usually rather inflexible and often requires several steps if the user wishes to change directions or otherwise exercise an option that is not immediately available at a given point of progress.

One of the more common circumstances is when a user has begun or completed one search and decides to begin a new one. In a number of OPACs, three separate steps are required to enter the new search statement under these circumstances. First, the user must return to the main menu. He or she must then select the type of search desired: author, title, subject, and so on. When the OPAC acknowledges this second step, the user enters the search term or terms, the third step. The limitation of options at any given juncture in a dialogue and the segmenting of dialogue into separate and distinct steps can actually be quite helpful for novice OPAC users, minimising ambiguity and reducing the risk of a user's becoming overwhelmed and confused by a plethora of alternatives and different rules and conventions. Such actions emphasize simple and logical progression, proceeding one step at a time.

For users who have little or no experience with computerized retrieval systems, who are unfamiliar with library catalogues, or who have very straightforward needs, such an approach can help them understand what is happening and gain greater confidence and skill in the use of an OPAC. For the user who has greater experience or more complex needs in using an online catalog, the degree to which an OPAC tends to segment the progression of a dialogue can be something of a nuisance. The necessity of carrying out two or three or even more interactive step to get to a logical nearby point in a dialogue can come to be seen as overly rigid, simplistic, and time consuming, leading to impatience and frustration on the part of the user. Thus, OPAC design is faced with the formidable task of attempting to accommodate both novice and experienced users, and both simple and sophisticated needs.

Some OPACs have tried to minimize this problem by building in at least a few optional capabilities that allow dialogue to be expedited, or more sophisticated features to be used, without compromising the availability of the simpler actions for those who are less experiences or who have less complex searching needs. Among some of the features designed to expedite dialogue, one is allow the user to "stack" commands, that is, to enter two or more commands in a single statement. For example, in an OPAC that normally requires two separate steps to enter search terms in a combined author/title search—one to enter an author's name and the other to enter the title—there might also be an optional procedure that enables the user to construct a single search statement containing both elements.

Another example of stacking commands, this one pertaining to commands for displaying bibliographic records, van be drawn from the MSUS/PALS online catalog. Like many OPACs, MSUS/PALS can display individual records in several deferent formats, each containing a different amount and arrangement of information about the same item. Although there is a "default" from that normally displays, a user also has the option of requesting a longer, more detailed version.

When a search statement has retrieved several records, the user can select any one of them to display in the default format by entering the command DI 3 for the third record in the set. When the default format display, the user can then enter another command to display the long version of the record. But there is also on option for these two steps to be combined in a single statement, DI 3 L, which will immediately display the third record in a set in its longer format. Although stacking commands enables a user to expedite dialogue by combining two or more steps that may otherwise be carried out separately, there are other, usually less powerful features available in many OPACs that can also reduce the amount of time and effort required of the user.

The most common is simply to allow a user to abbreviate some of the language required in entering commands. For example, one OPAC allows the command name FIND to be entered in either its complete from or abbreviated as F. A number of OPACs offer similar options; they do not really affect the course of progression, but they do give a user the opportunity to reduce the amount of keying effort required.

LINKING ONLINE RESSOURCES

In the past two or three years, e-journals have become the largest and fastest growing segment of the digital collections for most libraries. Collections that a few years ago numbered in the few hundreds of titles now number in the thousands, and the rate of growth continues to increase.In many ways, archiving and preserving e-journals will be dramatically different from what has been done for paper-based journals. In the paper era, there was large-scale redundancy in the storage of journals. Many different institutions collected the same titles. The copies of journals being saved for future generations were the same copies being read by the current generation of users.

Many of the things that helped maintain journals for the long term were not differentiated from what a library did to provide current services. Other than in the case of preservation microfilming and the odd instance of shared book storage facilities, there was little conscious coordination of preservation activities, and in fact a level of redundancy was expected and thought useful. The common service model for e-journals is quite different than that for paper journals. Most e-journal access is through a single delivery system maintained either by the publisher or its agent. There is little replication, and only a few institutions actually hold copies of journals locally. Libraries can fulfill their current service requirements without facing the issues involved in the preservation of the resources.

Further, in the digital realm the issues involved in day-to-day service are quite different from those involved in long-term preservation. The issue of long-term archiving and preservation of e-journal content has become one of increasing importance. Specifically because of archiving concerns, many research libraries continue to collect paper copies at the same time they pay for access to the electronic versions. This dual expense is not likely to be sustainable over time.

Publishers are finding that authors, editors, scholarly societies, and libraries frequently resist moving to electronic-only publication because of concern that long-term preservation and access to the electronic version is uncertain. Publishers and subscribers, and publishers and archives, have formal contractual relationships; does there need to be a formal relationship between archives and subscriber? An archive that keeps its content dark poses less of a threat of competition to the publishers with whom it is working. A dark archive will also be relieved from having to maintain a current user interface, with all of the bells and whistles that users have come to expect, and from the complex task of maintaining information on who has access to what content.

On the other hand, insuring that content that is never used remains sound and free from degradation will be challenging. Keeping records of who has the right to access what and implementing appropriate access control mechanisms that recognize differential rights to various archived objects would be a major operational challenge.What content is archived? At first hearing, most people assume that e-journal archiving is basically concerned with the content of journal articles. Indeed, while articles are the intellectual core of journals, in fact e-journals contain many other kinds of materials.

Some examples of commonly found content are:

— Editorial boards
— Rights and usage terms
— Copyright statements
— Journal descriptions
— Advertisements
— Reprint information
— Editorials
— Events lists
— Errata

— Conference announcements

— Various sorts of digital files related to individual articles (datasets, images, tables, videos, models, etc.).

Which of these content types need to be archived and preserved for the future? Some of these types of materials will pose issues for publishers. Not all of these items are controlled in publishers' asset management systems. Some are treated as ephemeral, "masthead" information and are simply handled as website content. When such information changes, the site is updated and earlier information is lost. For example, few if any e-journals provide a list of who was on the editorial board for an issue published a year or two ago.

Another difficult content type is advertisements. Advertisements are, of course, frequently not tied to any given issue, and they change over time with the business arrangements of the publisher. In some cases, advertisements are specific to certain populations, and what advertisements you see depend on who or where you are. Deciding what of all that is seen on e-journal sites today should be archived and maintained will require careful consideration by archives, publishers, and scholars.Should content be normalized? The variety of formats of digital objects in an archive will affect the cost and complexity of operation.

In order to control such complexity and cost, an archive might want to normalize deposited objects into a set of preferred formats whenever possible. Such normalization can happen at two levels:·

— *File formats.* An archive might prefer to store all raster images in TIFF, for instance, and convert JPEG or GIF images into that format. Controlling the number of file formats will reduce the complexity of format monitoring and migration.

— *Document formats.* Many e-journal publishers encode article content in SGML or XML (or plan to do so soon). Most publishers create their own DTD (or modify an existing DTD) to suit their specific needs and delivery platforms. An archive might choose to normalize all such marked-up documents into a common DTD, reducing the complexity of documentation, migration, and interface software.

Normalization and translation always involve the risk of information loss. In archiving there may well be a difficult trade-off between information loss and reduced complexity and cost of operation.

Electronic Journals

Many aspects of publishing are being transformed by the arrival of the World Wide Web and its facility to distribute information electronically. For journals, the transformation has barely begun. A large number of journals are now available in electronic form of some kind, but otherwise little has changed. Recently, one leading publisher reportedly described the growth of electronic journals based on the portable document format (PDF) as the "first frontier" facing journal publishers just a couple of years ago. The second frontier is the emergence of links, he went on to say. Judged by this measure alone, journals may have appeared on the Web but have yet to adapt to it.

Publishing may have changed, but readers probably experience little difference between Web-delivered journals and, in many cases, their print originals. New and better features that can be supported electronically are now being demanded by users, however, supported by librarians, as the price of acceptance of these new products. Of these new features, links are one of the most important. Since 1995, the Open Journal project has been applying original software tools and techniques to support flexible linking in e-journal applications, based on selected journals which were available electronically but which were are not all exclusively electronic.

In doing so, the project foretold the impact that links will have. Links are not a superficial feature of the Web, nor are they simple add-on features for e-journals. Links have the power to alter the character of journals fundamentally, most obviously in the development of "distributed publishing" in which users can find items of interest irrespective of the publisher. Ultimately, distributed publishing may transform the way in which individual documents are compiled by sharing components or "objects", figures say, from different sources, and by using network-based software processes, or services, to enhance presentation. Links are important for a number of reasons:

— For users, links provide faster, more direct access to more information.

— For librarians, links support more effective information retrieval, especially from large archives, and can help with identified user phenomena such as "successive search episodes".

— For publishers, links add value to works, but in this context links need to be applied in particular ways to make it easier to maintain and manage large numbers of documents.

In UK, the Open Journal project was launched with three main objectives:

— to provide immediate access to electronic versions of existing journals (noting that this was a time when there were few e-journals)

— to provide powerful hypermedia linking techniques

— to support faster access to information, including journals and other Web resources

Of these, the second and third were the most important. The

project did not intend to become a journal producer or publisher. These objectives were largely achieved, although with some qualifications which will be explored below. The project also expressed a longer-term vision of link-based journals: "For publishers the "journal" now becomes a set of links: substitute links for the glue of the paper journal". The project produced some extensive demonstrators, notably of citation linking, but it has not been possible yet to produce a convincing demonstrator of how journals might be fundamentally "re-packaged" in this way.

In practical terms, the Open Journal approach involves four components, respectively, linking, resources, formats and location. The link service used in the project was developed as a research tool at Southampton University and was based on technology that has been commercialised, under other names, for other applications. It quickly became apparent from project users that whatever you call the underlying linking tool, it is the functionality that counts, and that functionality must be presented to users as transparently as possible. The functionality, and the novel twist in this case, is that the links are added to documents that have been requested by users as they are downloaded. Up-to-date links are published within the document upon delivery rather than authored in typical HTML style because they are served by the link service separately from the document source and added on the way to the client via a proxy server. Resources of a wide variety of types, from primary journals to databases, were generously provided by a group of twelve publisher partners most of which supported the project throughout its full period and some of which continue to work with the project developers.

The linking tools will be compliant with the linking component of XML, an important new format for Web documents that emerged during the latter months of the project.

Full compliance will be formalised following publication of the XLink and XPointer standards. Recognising that it will never be possible to serve the information needs of users from single resources or single Web sites, a vital feature of this approach was that links could be applied to documents wherever they are on the Web—the distributed journal scenario again. Three demonstrator journals were produced in the subject areas of cognitive science, biology and computer science.

These areas and the core journal resources contained within the respective Open Journals were effectively defined by the interests of the research teams and the publishers that supported the project. Development was further informed by subject specialists, including researchers, teachers and librarians, in each of the areas. For the purposes of this analysis, however, it is more convenient to consider the demonstrator Open Journals from the viewpoint of the linking features they highlighted. The three main link types in this application can best be described as citation links, keyword links, and PDF links.

Citation links act on references contained within papers. Print journals are constrained to work backwards in time, citing earlier works; electronically, links can also point forward in time, to later papers that have cited the one being viewed, or even to later papers that have cited any of the references contained within that paper. The influences on a work and its own influence can thus be tracked. How completely a thread can be tracked in this way depends on the number of citation links that can be presented, which in turn depends on the scale of the data that can be linked.

Given the small size today of the electronic full-text archive, citation linking on a meaningful scale requires a secondary database of abstracts. The project was fortunate to be able to work with data provided by the Institute for Scientific Information (ISI) from its citation indexes. These indexes not only include abstracts from papers but the references too, and this is the basis of its distinctive services, now translated to the company's Web of Science. Working with a much smaller, but still substantial (500 MB), data set than Web of Science, the project demonstrated forward and backward linking within the secondary data but went further, extending the linking capability to remote but accessible full-text journals.

For those journals it was shown that references could be linked—where the data set allows; in this example the data set was not big enough for comprehensive linking—to the secondary data or, potentially, to other full-text journals independently of the established authoring and publication process. An important and successful component of citation linking was a software-based citation agent developed for use in the project. This agent recognises reference data within a downloaded paper, matches the citations against a pre-indexed database of abstracts, and links the references to entries in the database where matches are found. All of these actions are performed in "real time" as the requested paper is downloaded.

Conceived as an autonomous processor, the agent was partitioned as a "library" of functions to allow it to be integrated with other programs. In the project, this approach was used to enable the agent to be used with the link service and PDF software. Some other research projects are developing tools for citation linking. While the Open Journal project emphasised text recognition, matching, and linking, other projects are concerned with software agents that can find cited works on the Web, and improved search services and parsing of other document formats to build automatic and comprehensive citation indexes. Another approach to serving links separately from documents is Hyper-G, an electronic publishing package based on a Web server with an object-oriented distributed network database and a separate link database.

Hyper-G has been used in journal projects with Springer-Verlag and Academic Press. In one sense, serving links in this way is a rough-and-ready but practical method for implementing relationships, represented explicitly as links, between documents from different sources. This is a flexible approach but because no control is exercised over the documents, the links can be unstable and need to be rigorously mantained. At the other end of the spectrum, an ideal application-independent and stable way of identifying documents or their components might be the Digital Object Identifier.

Publishers to have prototyped applications represented in the the gallery of the International DOI Foundation include Academic Press, Elsevier, Springer-Verlag, and Wiley. The DOI is not yet an accepted standard, and to become so would require wide agreement. While links simply provide access to works, however, the DOI has a more demanding remit: "The intent of the (Association of American Publishers) AAP's Enabling Technologies Committee (which designed the DOI) was to support copyright protection, while ameliorating inconvenience to users, by supporting technology that promotes interoperability'. It is not clear that these objectives can be easily reconciled without compromising user access. There are also concerns about possible limitations, such as restricting organizations that are permitted to assign DOIs to "legitimate" publishers. Between the DOI and link serving is Hellman's proposal for the Scholarly Link Specification Framework (SLinkS) which applies DOI-like identifiers to documents from different publishers but controlled via an intermediary service.

It is already clear from a number of publishing arrangements that the electronic scholarly literature will be dominated by cross-linking on citations between different journals and services. ISI Links has been announced as its means of mediating citation linking between Web of Science, collaborating publishers, and subscribing institutions. Linking applications where links are applied between different journals and documents directly managed by a single publisher have been described for the BioMedNet service, HighWire Press and the Institute of Physics.

Some Web-based abstracts services enable third-party users to create links to entries in these sites. Keyword linking is more contentious than citation linking. Keywords are a means of classification that does not automatically translate to linking on the same scale as citation linking. Keywords are often created by the authors of papers, and therefore may not always be compiled systematically. Publisher and editor intervention can help by formalising the terms selected, especially in structuring the terms in a hierarchy.

As a result, however, indiscriminate linking based on keywords can produce results which appear not to match authorial intent and produced, early in the project's first demonstrations, some adverse reactions from users. Given the ubiquity of keywords within quality journals, and the frequency with which terms might appear given their position in the classification hierarchy, it became a relatively simple task programmatically to develop and display large numbers of keyword links. This creates problems for users, in this case the classic problem of information overload through too many links. Links appear to be random and are not well labelled; in other words, users are unsure where a link will take them or whether following the link will be useful.

Links to dictionaries or glossaries produced from keywords can be more intuitive and useful, but even here the response was equivocal. Dictionaries are often assumed to be low-level texts, useful for novices and students. One way forward, instead of using ready keywords, is to look again at texts for the occurrence of what might be called link words, quite a different concept from keywords and which are created with a better understanding of the linking strategy.

With new demands being placed on e-journals, perhaps link words will become an editorial task as common as creating keywords today. On a smaller scale, it would be possible to use this technique effectively within single journals, where the effect of keyword linking would be to overlay the journal index as links on the electronic text archive. PDF linking is different again, concerned with the technical issue of managing data in a particular format rather than the content of that data. Of course, the content could be a citation link or a keyword link.

Most electronic journals are in PDF so it was natural for the project to treat this as an important linking issue. Although Adobe designed PDF to include links, links are not a primary feature of the format as they are within HTML, and it is notable that few PDF journals have links. Among the few that have citation links, some such as the American Physical Society and the Institute of Physics in the UK have reformatted reference data into HTML for the purpose.

There may be good reason for this. PDF may be chosen for ease of conversion from print sources, but once in that form the data is complex and difficult to manipulate in the way required by the project's approach to linking on a large-scale, an approach which is more sophisticated than linking supported intrinsically by the format. It required a major effort, but the project produced a working service for linking from PDF. It remains to be seen whether the dominance of PDF prevails for e-journals, but since it is more cost-effective to work with given formats rather than convert then PDF linking could be an important tool. Although converting references to HTML is one alternative, it does not address the need for links, keyword links perhaps when the implementation is refined, within the body of papers.

There is, however, the danger that the PDF tools will have to be updated every time there is a change to the linking framework, as happened during the project, or each time Adobe changes the specification for PDF or the way in which it supports the format. In principle, because Open Journal linking is a server-side process, this approach is both independent of the platform used and of the version of a given software application on the user's machine. Adobe PDF may be a de facto standard as far as e-journals are concerned, but it is still a proprietary format, and this is not ideal in an environment such as the Web which promotes the use of open, public standards that are intended to allow improved interoperability between applications.

User responses were gathered by the project in a number of ways: meetings with individual users, selected evaluation groups, and open access demonstrators, mediated by electronic mail and Web forms; and less directly through meetings with other interested groups, conference presentations, seminars, and workshops. Evaluators included specialists in the subject areas covered, as well as teachers, librarians and publishers. Important results from user testing

— Users are very demanding.

— "It would have to be much faster."

— "Speed of links -- initial link to an abstract, have all the abstracts loaded with the initial page"

— Citation links are popular, and users will always want more links.

When services are simple, intuitive and useful, it is hard to overstate the impact, even for new users. The Web as a whole is one example. Citation linking will be another case in point. With the exception of Acrobat it seems the practice of software download does not apply to typical e-journal users, as the project soon discovered from publishers and librarians. So the link service software was rewritten to work at the server end, mediated by a user-set browser proxy. Even this was insufficient.

Libraries do not want settings on shared machines to be altered, and proxy settings can interfere with firewalls in corporate environments. In the latest version the direction to the proxy server is attached to the URL, leaving the user to browse the Web conventionally and do nothing to receive the link service, barring starting from the right place! The interface is the user's starting point, and Web interface design has advanced enormously in the last year or two, especially for large resources, where so-called "gateways" have leapt in importance. An Open Journal needs not just links but a distinctive entrance, or gateway, too.

Distributed Link Service: A connection Source to link server

DLS allows the client to connect to a link server to request a set of links to apply to the data in a document. From an abstract viewpoint it provides a hypermedia link service which can be used alongside, but separately from, the WWW's document data service: in practice the link service is mediated by the WWW, and implemented by CGI processes located on Web servers. The WWW has a well-established method for expressing links as attributes of its native document format, and so the link service will provide a complementary set of links on top of those standard facilities.

By contrast, a simple text editor has no built-in hypertext links, and so the link service provides an otherwise non-existent service to such users. Without a link service, Web users can follow links from HTML documents or 'imagemapped' pictures into dead-end media such as spreadsheets, CAD documents or text; with the link service they can also follow links out of these media again. A link service can provide several practical benefits to information systems: it can act as an adjunct to provide hypermedia facilities for applications without requiring a rewrite of the applications themselves, and, by allowing links to be manipulated as independent objects can aid in the document maintenance effort since a document does no longer needs to be revised in order to change its links.

In fact, a document does not even need to be revisable; links may be applied to legacy documents or data stored on read-only media, and the relationship of a document to the larger docuverse may evolve over time, even though the document itself remains static. A link is often considered to be similar to a GOTO instruction in a programming language, and since it is useless to examine a jump instruction extracted from its program context it would seem that a link extracted from its document would equally be an oddity.

However, a link can instead be seen as a specification of a relationship between two data items; it expresses a relationship between its source and destination, and may even be expressed in a way to allow its source or destination to be parameterised. In this way the source may expand to one of several offsets in a particular set of documents, or the destination may resolve to any of a number of alternative documents. These flexible relationships can be expressed in a simple way, and stored together in database; the link service is effectively a database lookup service where the data items are interpreted as links between other data items. The HyTime standard for hypermedia links provides a comprehensive set of mechanisms for coding the relationship between objects using an SGML representation.

It allows for both HTML-style links, where the link is coded in situ at one of its end points, and also for independent links, where the link is coded in a section or document separate from the data which it links together. Although the DLS does not yet use the HyTime standard coding, it uses the HyTime concepts of links as independent entities. The links themselves are distributed across a number of databases and the databases can be distributed across a number of servers. Further, the document server may provide only a basic set of link databases, whereas the end user may have a private set of link databases and a third party server may provide some sets of specialist links on a commercial basis.

Users of the DLS

Given the existence of this link service, there are a number of different ways that it can be used to achieve benefit for different groups of users.

Readers or Browsers

The end-users (readers or browsers) may choose to subscribe to this service by running a small interface agent which communicates with both the link service and the document viewer. For an information consumer on the Web, the link service provides an additional means of navigation that can be tailored very precisely to his or her exact needs. When the user wishes to investigate links from some information, they select the data of interest and choose the Follow Link menu item from the interface agent. The agent grabs the current selection, tries to determine the current document context, parcels this information into a message which is sent to the link server.

This process consists of creating an HTTP message with POST data and sending it to a Web server, since the link service is actually hosted by the Web. The link server then responds with a set of links which are available from the specified selection in the specified document. These links are presented to the user in the form of a 'clickable' list of destinations, displayed as a page of HTML by the Web viewer. The user can select the Show Links menu item instead. This will send a similar message to the server which causes the selection to be analysed word-by-word or phrase-by-phrase. All links which apply to any of the constituent words or phrases will be returned as if the user had selected Follow Link on each of them. In some respects this process is similar to making an enquiry of a Virtual Library—certainly the same kind of database lookup is going on, and a similar-looking page of HTML is returned.

On the other hand, this activity is both document centered, occurring while the user's focus is on the information content instead of the navigation superstructure, and context specific, parameterised by the location of the query data. If this action is not totally like a database query, it is also not completely like a normal link following operation. It is reader-driven rather than author driven as it is the user who decides what is of interest for link following rather than the author's sole perogative to determine what material can be followed up. This has some positive benefits for some application environments where an exploratory style is encouraged, but may be less useful for other areas where a more prescriptive style is appropriate. The DLS is a particular kind of information service, and so bears some similarity to a WAIS server.

In fact one kind of hypertext can be implemented using just such a text retrieval service, where the links are all derived by a statistical analysis of the lexical data contained in the documents. This style of hypertext, known as a hyperbase, is characterised by the lack of internal structure as opposed to a hyperdocument in which the links are created by human authorship. Services like WAIS and the Virtual Libraries provide some similar features to the DLS, but lack the capability to express human-authored links.

The DLS on the other hand can implement a range of links, from the human authored variety through to the automatic keyword matching variety. Another feature of the DLS is that it allows link following to be affected not just by the static document context of the link source, but also the reader's dynamic context: task-specific information that affects the kind of resources that they would like to follow up. A reader may choose to subscribe to a subset of many different linkbases, depending for example, on whether they are to write an essay for an undergraduate or postgraduate course on Cell Motility, or whether they are interested in Computer Science in general or Parallel Computer Architectures in particular.

This is achieved by the interface agent querying its link server about the link databases it supports; the list of named databases is offered to the user as a Context menu. By selecting a different set of links using the Context menu, the user will see different responses to the Follow Link request: in effect they have the ability to reconfigure their hyperdocument, controlling their own view on the web. This can also be further achieved by choosing which link servers to connect to: a user may improve their Web connectivity by subscribing to a new commercial host which offers an enhanced sets of links across a greater set of Web sites.

Authors

Authors may make use of the DLS by using the same interface agent. Since a part of the authoring process involves the author taking on the role of a reader, the author can benefit from the link service exactly as a reader can, but in addition an author can create links and edit link databases. To create a link, the author highlights the link source, chooses Start Link from the interface agent, finds the link destination, highlights it and chooses End Link. At this stage the interface agent packs the selections, static document context and dynamic user context information into a Create Link message and sends it to the link server via the Web.

The newly-created link may be added to the user's personal link database or to the current Context database; either of these may be subsequently edited by a forms-based interface in the Web browser. This kind of functionality is fairly straight-forward, but the real advantage for the author comes in the kinds of link definition that are allowed. Following the Microcosm model, upon which much of this work is based, links may be declared to be more or less generic, i.e. having the location of the selected text constrained to appear more or less specifically within the static document context. A standard link applies only at the exact place that the link source was selected, whereas a completely generic link will match the link source's selection at any place in any document.

This facility allows the author to treat a link as a declaration which states "any place in such-and-such a document context that phrase 'X' is mentioned links to this data", and allows the author to create a set of documents along with a set of links that can be used to 'come to' the documents from other places as well as a set of links which 'go to' other documents from the current documents. The 'come-to' link type leads to a resource-based authoring style in which an author can publish a largely standalone suite of documents, together with some link databases which define the 'routes' into, through and out of the documents.

Making use of the link service allows the author to 'mix together' a number of these resources as the 'into' links for each of them will act on the text of the others and bind them all together. In fact, the 'into' links can act to bind the resources not just to each other, but to the larger Web of documents outside the author's control—the readers' environment. One of the major benefits of this authoring style is the scope for information reuse: not only can the author vary the internal paths through the documents by changing the link databases, but also the documents themselves can be used and reused in many different situations by providing different sets of 'into' and 'out of' links.

Link databases use an SGML-based format, and so can be processed using standard tools like Perl, sed or awk. In particular it is possible for an author to automatically create a simple link database from scratch by extracting sets of keywords from documents and turning them into generic links. Such a database would need a certain amount of author-editing to become properly useful, but seeding a set of links in this way can provide a basic amount of hypertext functionality with minimal effort.

Publishers

Publishers, information providers, or information resellers may make indirect use of the link service by providing link databases for their customers to access in their roles as readers or authors. Publishers may in fact be ideally suited to act as link brokers, providing a more centralised service than hundreds of individual authors themselves. There is however an alternative role for the link service in the publishing environment as a tool for the management and generation of HTML resources.

As a variation to the standard use of the link service, a user can choose to send to the server not just a selection from a document but an entire document. The link service will then undertake to return that document with all the applicable links hard wired into the document in whichever native format the document was created. This option, available from the Compile Links menu item on the interface agent, currently handles HTML, RTF, PDF and text document. The effect of this facility is to place the links belonging to a document back into the document so that the new document can be used independently of the link service, especially if the document is to be distributed as part of a non-networked environment.

This facility also allows publishers to work with a resource-based paradigm, creating a Web of information nodes for a specific niche market, and recreating a different Web over some of the same information for a different market. The Open Journal Framework project funded by HEFCE's Electronic Libraries programme in the UK is capitalising on the link service's functionality from the publisher's points of view. The aim of the OJF is to enable electronic journals in a digital library to become co-operating assets in an information delivery environment rather than isolated, one-off resources.

Based on a set of several hundred PDF files, which correspond to several volumes of the Company of Biologists' journal Development, work is being done to establish link databases for those resources along with other network-accessible biology resources. The OJF can easily be extended to provide guided or free searching through a prepared database to create a powerful learning environment. Installing expert system technology into the link server allows it to become an intelligent tutor to direct students through material available on the network. The Company of Biologists is at present negotiating with other publishers to allow the use of standard texts to complement the content of their journals for teaching purposes. The benefits of the project therefore extend beyond electronic libraries for research and into customised teaching and learning environments.

Implementation of the Distributed Link Service

There are two aspects to the implementation of the Distributed Link Service: the client side interface tool, and the server-side link database utilities. All communication between the client and server is provided using standard WWW protocols. The DLS client uses a standard WWW client to send linking requests encoded as URLs. These URLs allow the client to access a link server which operates via a WWW server.

— *Link Server:* The server facilities of the DLS are implemented as CGI scripts, and are accessed using a standard WWW server. The main scripts are those that allow the creation, following and editing of links which are stored in link databases.

— *Server interface:* To carry out their functions, these scripts have access to a variety of different link databases, the exact selection depending on the end users configuration.

There is a main link database for the server, which is always used, and additional link databases from which the user may choose. These additional databases allow the server to offer a range of different link sets, known as contexts. The end user is able to select a context from the whole set, thus allowing their linking activities to be tailored to their current requirements. For example, different contexts could be provided to cater for different levels of user knowledge.

In addition users are allocated a personal link database in which any links they wish to maintain for their own use may be stored. Databases are stored using an SGML style markup, and record details of the source and destination of a link, the link type, time of creation, and the description of the link. The various scripts determine the appropriate context from the information passed to them by the end user's WWW browser, and the user linkbase from the details of the user and host connecting to the server. The scripts then carry out their activities using the appropriate databases. The script which handles requests to follow links simply checks in all appropriate databases for any applicable links and returns these to the user as a list of potential links in an automatically generated HTML document. The user may then go directly to the link destinations.

The show links request is a special case of the follow link process, it takes a large selection and carries out several follow link requests on the words and phrases making up the selection.

The script which creates links accepts details of start and end points for a link, and inserts a link into either the database for the specified context, or the user's personal database if no context is given. The edit link script offers a fill in form which allows links to be entered directly into databases, deleted, or their details modified.

In particular the user is able to update the description of the link, which by default is the name of the destination, or change the type of the link to make it more generally available. Links may be defined as one of two types: local, in which case it applies only in the document in which it was originally authored, or generic, in which case it applies whenever the source anchor selection is found, thus reducing authoring effort significantly. Finally, a script is available to compile links held in these link databases into an HTML document, thus providing direct access to all applicable links without the need to make individual requests to the server. This can also provide a way in which material may be authored using the DLS and delivered as standard HTML.

Link databases: Link databases are the essential components of the Distributed Link Service.

A link is of type specific if its source anchor is constructed from a complete triple (i.e. a specific occurrence of a selected object in the named document), local if its anchor is ignores the offset component of the triple (i.e. any occurrence of the selected object in the named document) or generic if only the selection is used (i.e. any occurrence of the selected object anywhere in any document).

In the Microcosm system (upon which this work is based) offsets are frequently used, but because of a lack of technical integration with the various viewing programs, the DLS usually ignores the offset. Hence only local and generic links can be created and manipulated by the various user interfaces, although it is possible to process specific links programmatically. Note that the DLS provides flexibility in specifying the source anchor: this means that a single link to a destination may appear in many places at once, giving rise to a number of useful features for the hypertext author.

Linkbase processing: When a Web server hosting the DLS software receives a followlink request for a selection in a document D, the followlink script gathers together a number of link databases to try to satisfy the request. These are:

— *A document-specific linkbase* containing links which are pertinent only to that document itself. This linkbase is stored as a file in the document subdirectory of the DLS software.

— *A resource-specific linkbase* containing links which are pertinent to a group of files (probably a whole directory hierarchy). This linkbase is also stored as a file in the document subdirectory of the DLS software.

— *A server-specific linkbase* containing links which may be relevant to any files provided by the server. This linkbase is stored in the file server.links at the top level of the links software installation.

— *A user-specific linkbase* containing links created by the user who issued the request. This linkbase is stored as the file user@host in the user subdirectory of the DLS software installation.

— *A context-specific linkbase* containing links customised for a specific task which the user interface allows the reader to choose at any time during a session.

This 'chain' of linkbases is examined when the user asks to follow, show or compile links. When a user makes a link, the link is either stored in their personal linkbase or, if they have a chosen a context, in the context database. Currently there is no way to create a link in the document, resource or server-specific linkbases: these must be created in a dummy context and manually transferred at the server.

Client Interface : A client interface is available for PC, Mac and UNIX platforms, and is a fairly simple utility whose task is to formulate requests for linking functions and to dispatch these requests to a DLS link server for processing.

The main task of the client utility is to react to a request from a user to carry out a particular linking function by extracting the details of the selection the user has made, and if possible, the details of the document in which the selection is. This information is encoded in an HTTP request reflecting the chosen function, and the selection details.

This request must then be delivered to the currently specified link server. This is done by communicating with the user's WWW browser, and requesting it to load the new URL. The results of the link request are returned as HTML, and displayed by the WWW browser. For example, in response to a request to follow a link from a particular selection, the response might indicate that no links were found, or list the links that were identified. All versions of the client interface are currently designed to support the use of Netscape 1.1 for communication with DLS servers.

The nature of the client interface, and the way in which it communicates depend on the particular platform being used. In the future, some or all of the functionality of the client interface may be achieved through programmable features in browsers. The first version of the Distributed Link Service was released in early June 1995, initially to the ERCIM (European Research Consortium for Informatics and applied Mathematics) WWW Working Group (W4G). Another major project is the Open Journals Framework, which uses the software to support electronic publishing. Two other major projects involve a traditional Web server for large commercial organisations, and use the DLS in conjunction with our HTML 'compilation' tool to ease the significant information maintenance and authoring task.

Other projects include a Web-based distance learning course, which adopts a resource-based approach, and, in contrast to this manually authored service, the support of an information service for university administration: different users are provided with different views of the data according to their role, and the link service is used to automatically link new documents. In each of these applications, the link databases are held on the same server as the documents. This is the simplest use of the DLS and is adequate for many applications. Meanwhile, the fully distributed version of the DLS is under development: this supports linkbases maintained on different servers, enabling the linkbases to scale and supporting localisation of linkbases. Localisation means that an individual user, or a group of users, can work with databases held as locally as possible.

An early prototype of the fully distributed service used SMTP to provide a standard asynchronous messaging model for inter-server communication, with MIME-encoded link resolution messages. Taking a query-based approach can cause a significant loss of performance where users make use of a proxy, since the results of the queries are not cached. This problem arises when there are remote users of a single DLS server, especially if those users are widely distributed. There are various solutions: provision of a proxy with DLS extensions would allow require remote users to change their proxy, so the preferred solution is to modify the DLS server so that intermediate proxies cache appropriately.

Subject Gateways

A subject gateway, in the context of network-based resource access, can be defined as some facility that allows easier access to network-based resources in a defined subject area. The simplest types of subject gateways are sets of Web pages containing lists of links to resources. Some gateways index their lists of links and provide a simple search facility. More advanced gateways offer a much enhanced service via a system consisting of a resource database and various indexes, which can be searched and/or browsed through a Web-based interface.

Each entry in the database contains information about a network-based resource, such as a Web page, Web site, mailing list or document. Entries are usually created by a cataloguer manually identifying a suitable resource, describing the resource using a template, and submitting the template to the database for indexing. Subject gateways are also known as subject-based information gateways (SBIGs), subject-based gateways, subject index gateways, virtual libraries, clearing houses, subject trees, pathfinders and other variations thereof. There are a considerable number of Web-based gateways that can be used to locate network-based resources in some particular subject area. Nearly all of these gateways have unique features, additional subject-based services, and different approaches to how information about network-based resources is stored in the resource description database.

Most subject gateways allow the end-user to either search or browse the database of resource descriptions. For example, the AstroWeb Astronomy gateway consists of a browsable multi-level menu of sub-areas and resources, as well as a WAIS-based search mechanism. In addition, most gateways allow the user the options of case sensitive searching and stemming, where resource descriptions containing variations of a term are located. Some gateways provide extra facilities for enhanced searching. SOSIG, the Social Science Information Gateway, incorporates a thesaurus containing social science terminology. This gives users the option of generating alternative terms/keywords with which to search the resource catalogue.

SOSIG also allows users to search on resources that are located in distinct geographic areas, such as in the whole world, just in Europe or just in the UK. EEVL, the Edinburgh Engineering Virtual Library, allows users to search on a subset of types of resources, such as electronic journals, mailing lists and/or conference announcements. PAW, the Physics Around the World gateway, allows users to select the level of suitability for educational physics resources e.g. kids, school or university. Aqueous, a gateway dedicated to water related Web sites, allows two levels of searching. Some subject gateways have built a considerable number of related services and information sources around their core searchable/browsable gateway.

Biz/ed, the Business and Economics information gateway for students, teachers and lecturers, contains large amounts of business and economic information, such as company financial data and economic datasets from sources such as the office of National Statistics. Eldis, the Electronic Development and Environment Information Service provides access to several related bibliographic databases. History, a gateway to network-based history resources, provides details of historians and their research and teaching interests. OMNI, a medical and health subject gateway, has a section which allows users to purchase health-related CD-ROMS.

Citation Linking of Online Journals

The number of online journals is growing dramatically is becoming something of a cliché, but not everything is changing so fast. As established publishers bring existing journals to the World Wide Web we are witnessing not so much an evolution but, in many cases, direct replication, even down to the actual appearance of the original printed page when it is viewed online. If the journal article is the basic building block of the scholarly literature, then the journals are the walls. By leaving these walls in exactly the same arrangement as in the print environment, however, the limitations of which are becoming increasingly apparent, the online user is just as enclosed in his or her subject domain, within his or her library, as before. Simply, this is not sustainable. By embracing the Web, as journal publishers are, the assumption must be that everything, not just certain things, will change.

This includes the structure of journals publishing and the journals themselves, even down to the form of individual articles and the ways and styles in which they are authored. The most fundamental problem facing journal readers, library users and, it follows, librarians, is physically to get hold of all the journal articles they need when they need them. Experience varies enormously of course, but the days in which every essential journal was held in all major academic libraries have passed. The reason most commonly cited is economic, that journal price inflation has outstripped library (and personal subscriber) budgets, with rationalisation the inevitable result.

There has also been growth in the number of journal titles, which reflects changes within traditional academic disciplines: broadening and subsequent fragmentation within fields, and greater emphasis on cross-disciplinary work. With the emergence of the Web, conventional, largely failing, print-world responses to these issues may no longer be relevant. Cutting through this complexity, some points stand out. Rightly or wrongly, the dominant motivation for scholarly journals publishing today, whether by commercial publishers, learned societies or university presses, is profitability.

Commercial publishing exists to serve markets, so users, librarians too, have an influence. The above suggests that journals markets are not being served, however. While the long-running arguments over journal prices, the so-called 'serials crisis', have failed to resolve this problem, indeed have exacerbated it, the likelihood is that this issue of access will be a turning point. It affects most users more directly than do journal prices. The Los Alamos physics e-print archive, for example, is successful not because it gives access to papers more cost-effectively than do print journals (which it does), but because it fulfils a latent need that was not otherwise being served by publishers or which could not have been served by earlier technology.

The critical new factor is the Web, a massively popular platform that not only offers the chance to prevent further deterioration, but to transform access to the literature. The technology of the Web can be used for different purposes to suit different users in different disciplines, but as soon as that new functionality is fully realised so the purpose of journals publishing in all disciplines will also, ultimately, be transformed.

In principle academic research is progressive, one of its central tenets being that work should not be unnecessarily duplicated. Naturally this is reflected in the structure of the published literature: originality of work is a prerequisite for submission to journals, and the published works themselves form a record that informs, and might even direct, later research. Scholarly articles are not intended to be used in isolation, and one way in which this is reflected is the ubiquity of reference lists within modern papers.

Although not referring specifically to scholarly works, Barlow adds a new perspective: 'Information that isn't moving ceases to exist as anything but potential ... at least until it is allowed to move again.' For a scholarly work, what characterises 'moving' is its ability to influence other work, that is, to be read, commented, acted upon or referenced by others in the field. While it can never be guaranteed that a work will be acted on or responded to, its influence will certainly be seriously proscribed if reader access to the publishing journals is difficult. Networked electronic, or online, publishing represents a profound change from existing practice that has become commonly delineated in this way as a switch from static (print) to dynamic or 'moving' (online) information.

Okerson anticipated the significance: '... the moment information becomes mobile, rather than static, this transformation fundamentally alters the way in which information is used, shared, and eventually created.' This is not universally accepted. Peek and Barron, for example, are more cautious about claims for technological change and its impact, but in this case the evidence for sustainable change-current experience and the widespread acceptance of the Web-is compelling. There are persuasive historical visions.

Licklider and Taylor believed even then that they were 'entering a technological age in which we will be able to interact with the richness of living information-not merely in the passive way that we have become accustomed to using books and libraries.' The key to realising a dynamic journal literature is integration of information resources. It is not through the imperative of this cultural shift that integration is on the agenda, but basic market forces. The revelation of the Web, to commercial publishers at least, is that they can no longer be sure what has value, or what people are prepared to pay for.

The volume of material freely available on the Web, and its improving quality in parts, suggests that it is not textual content that obtains financial value automatically as it does in print. Instead, it is widely suggested, services are the key. According to a European Commission study on electronic publishing developments, 'electronic publishers must focus on integrating their content with services, customer-driven product tailoring (and) the brokering of information.' When print articles and journal issues were simply discrete entities, libraries and secondary publishers provided some integration through services such as bibliographic cataloguing, indexing and abstracting. The introduction of Web-based library catalogues will begin to speed up access to individual journal articles where electronic versions are available, but there are still many problems.

Critics such as Dempsey have highlighted the gap in providing detail about journal articles actually held by an individual library. On the other hand, digital libraries such as the Perseus Project have shown that a critical mass of multimedia, multivalent materials can be a catalyst for the transformation of practice. Indirectly, there is evidence that users are ready to accept dramatic changes in practice. A major stimulus for improved access to electronic articles is increased electronic collaboration between researchers. Wulf defined this environment as the 'collaboratory', merging 'collaboration' and 'laboratory' to describe a '.. center without walls, in which researchers can perform their research without regard to geographical location-interacting with colleagues, accessing instrumentation, sharing data and computational resource, and accessing information in digital libraries.'

The recently completed Distributed, Collaboratory Experiment Environments (DCEE) in the USA involved seven testbed projects representing scientific environments of various scales and involving from tens to hundreds of scientists and students. Search engines are widely used on the Web, and increasing use is being made of custom alerting services to bring newly published material to attention promptly. To these means of integration must be added the hypertext link. The link is the single most important, characteristic feature of Web pages. Just as it is rare to find links in online journal papers, especially those reproduced from a print original, so too citations were absent from the earliest journals.

Given that citations are an integral feature of the modern

scholarly paper, it is not hard to anticipate how citation links will proliferate in Web-based journals. The likely effect of citation linking can be gauged by recognising a direct parallel with citation indexing, which is sometimes referred to as 'forward' referencing (i.e. for a given article, all the subsequent papers that cite it), developed by Garfield. Citation linking combines the two approaches, mapping both reference data and citation index data on to the text in the form of links.

Adding electronic links to the literature is introducing a new culture in many respects, but citation linking is likely to be acceptable to the academic community because it builds on practices established in other forms such as print. Further, citation linking between journal articles will only be effective if links are direct to the referenced article, and this immediately presents difficulties, not just technical difficulties, but the need to pass through commercial walls, between other journals or other publishers, that are a legacy of print journals. New applications are highlighting this problem, but interim solutions are emerging. Another example, Physical Review Online Archives (PROLA) posts two lists on each document information page: the first provides a link to all PROLA articles that are referenced by the current article, and the second lists links to all the later papers that reference the current paper, including any errata that may have appeared.

'This hyperlinked cross-reference system allows users to locate and traverse a thread of related articles.' PROLA offers hyperlinks to articles within the PROLA system, will automatically locate and list all references to archived articles from the 1996 online APS journals, and there are plans to offer links to other, non-APS, online journals that reference Physical Review.

Institute of Physics Publishing (IoPP) in the UK was one of the first journal publishers to make all of its titles available online. Its latest development is the Hypercite Project, through which a subscriber can link from a reference in any online IoPP paper to an abstract (if it exists) from an Inspec-provided database or can link directly to the full-text article within another IoPP journal. Online versions of IoPP journals can be purchased as a site licence on terms, in the UK at least, which encourage purchase of the whole journals catalogue, thereby making it possible to link between full-text papers within IoPP's site.

The wider world of full-text physics papers beyond IoPP is mimicked by substituting the abstracts service. For users, even this should prove to be a significant step forward. Similarly, Electronic Press, through its BioMedNet service for scientists in biology and medicine, links citations to full-text articles in journals it owns, and other citations are linked to abstracts in its Evaluated Medline database. HighWire Press, a producer of 'high impact' biomedical and science online journals, is another to have begun exploring the use of links between full-text articles available through its services or to abstracts in a Medline database.

Broadly, the project, funded by the UK Electronic Libraries (eLib) programme, addresses the issue of information coherence in the information environment currently served by academic libraries. An increasing proportion of the diverse information assets in this environment are becoming electronic, and from a single 'terminal' users can access journals, databases and articles but, as we have seen, are required to navigate a complicated path through many providers' information gateways to locate any particular piece of information.

The goal of the project is to develop a framework of information retrieval technologies (based around the DLS) and to establish electronic publishing practices to be used by information providers (especially journal publishers, but also by librarians, perhaps) which will allow them to make publications available not as isolated resources, but as cooperating assets within an information delivery environment: the 'open' journal. The facilities of the DLS, which allow authors to create links between arbitrary Web documents and then to publish those links for the benefit of other users, provide a basis for reintegrating the online literature.

In hypertext terms, they allow authors to turn a bounded subset of the Web from a hyperbase, an undistinguished database of information nodes, into a hyperdocument, a collection of documents united by authorial intent. One of the major features of the DLS is that, instead of inserting link data within the html (or other) coding of a source document, the data are held separately in a link database, or linkbase. The links can be viewed superimposed on a Web page, as though a transparency were overlaid on it, by retrieving the document from a url via a proxy link server. This is easily set up using the standard menu facilities on popular Web browsers.

The links appear to act as conventional Web links, so the application of the link service is effectively transparent, while at the same time it provides the link author and user with significantly improved link flexibility. It is this capability of separating links from the authored content which is perhaps the key to allowing journals to interoperate, because it allows the creation of links between two third-party resources, not just to another publisher's documents, but from them as well. Imagine an institutional library has introduced a new rule governing the use of its books and journals.

To be allowed to take out a book, users need to present the issue desk not with the book itself, but some other book which contains a citation for the book actually required. Only when the librarian sees a valid citation for a book or journal will the user be allowed to read it. Further, the staff choose to keep all books behind the counter, inaccessible to library users but directly available to qualified librarians. In this scenario there is no way to idly look through stacks of new periodicals to keep abreast of the latest developments, nor can users select the particular book required on a topic by inspecting the whole range of books available on that topic.

This is almost the situation that users of the Web find themselves in, an environment where a private collection of documents can be requested individually, but only given a valid reference to the document. In both situations, neither the identity of a document, nor even its existence, can be deduced or guessed. Under the new rules there are two approaches that will enable users to have access to all the related information they want from a collection of disparate books and journals from a range of publishers:

— produce comprehensive 'meta-books'

— provide links leading from every article, book and journal

Much activity has been directed at the former-producing lists of content, and lists of lists, making catalogues and indexes of subject material on the Web-but less in authoring and publishing links, towards which the link service is directed.

Despite the above examples of citation linking, the scarcity of links in the majority of online versions of established journals is partly a reflection of this situation. One likely effect of using a link service in a professional publishing environment, such as journals publishing, will be to transform link authoring into a link publishing task, possibly handled by link editors. In this way links are added at the moment of publication rather than during the authoring stage, an important feature in a dynamically changing environment such as the Web.

The link maintenance problem also becomes more manageable. In addition the Open Journal project has extended the core functions of the DLS to work not just with html documents but also portable document formats viewed with Adobe Acrobat. To demonstrate the concept the project is producing three Open Journals in the areas of biology, cognitive science and computer science.

Each consists of original journals from a number of different primary and secondary publishers, served from a number of different sites in a number of data formats (principally html and pdf). Dictated by differing cultures, academic disciplines show various degrees of acceptance of online publishing, adopting different approaches. This is borne out by clear distinctions between the three Open Journals. Biologists have access to a diverse range of online materials, from journals to books, databases of gene sequences and other molecular structures, libraries of graphical images and teaching resources. The information is invariably well catalogued and often freely accessible.

The databases in particular form a focal point for the field, lending it an online coherence that few other disciplines can match and attracting a higher proportion of the field's prestigious journals online than other fields. This area also provides the project with its longest list of participating journals, one of the project objectives in this case to begin to link these journals into the non-journal data resources. In contrast, computer science offers large volumes of disparate online resources with little organisation apart from some loosely catalogued indexes, although there is an attempt to build a coherent 'library' of networked computer science technical reports. Few established computer science journals are available online, and the odd new online journals in this field, with one or two exceptions, have been noticeably slow to grow.

In this case the project aims to identify pockets of coherence that can be extended through the application of links, starting with specialised journals provided by publishers collaborating in the project. Cognitive science, likewise, seems ill-served with online journals. It is ironic, then, that the field delivered one of the first online journals of all, Psycoloquy, edited by Stevan Harnad and launched in 1990. This journal and its extensive archives, together with its print counterpart Behavioural and Brain Sciences, which recently began an online preprints service, are the nucleus of the project's cognitive science Open Journal. As has been argued, most journal papers available on the Web are unlinked. The DLS can be used to overcome the lack of links on most Web journals by dynamically matching a fixed database of pre-defined links against the contents of an article, inserting a link whenever an explicit match occurs.

More general than a programming language GOTO statement, a link is the specification of a relationship between a data source and destination where both the source and destination may expand to one of several places in a particular set of documents. These flexible relationships are usually coded explicitly by the DLS as a generic link, but they may be the by-product of a more complicated processing arrangement. For the purpose of citation linking this involves matching citations and bibliography lists in a document against a database of bibliographic information. This is a complex process which requires an understanding of the house style of the publication, an understanding of the document coding format and the ability to parse complex language structures.

Since this is a knowledge-based activity being undertaken on behalf of the user, we call it a citation agent. When the DLS processes a document which is defined to be in the Cognitive Science Open Journal, it launches the agent to act on the document before adding other links in the usual way. In short, with the arrival and, in some fields, broad acceptance of a core mass of scholarly online journals, the most progressive of these journals are evidently heading towards a new phase in which they are being enhanced to take advantage of the capabilities of the Web. Tools to support this new functionality are proliferating. Among them, link services offer the potential to transform the use of the dominating feature of the Web, the hypertext link.

3

Digital library Cataloguing Systems

The principle activity of cataloguing systems remains the production of catalogue cards, but the availability of high-speed printers driven by computers has sparked a remarkable revival of interest in an older way of presenting cataloguing information, the book catalogue. Other printed products, created usually as by products rather than as a primary goal, include spine labels, book plates, book pockets (or labels for placement on book pockets), and book cards for use in either manual or automated circulation systems.

In recent years, there has also been a growing recognition that substantial economic benefits would result if cataloguing data could be provided to individuals libraries in a timely fashion from a centralised source, thus lowering the cost of the intellectual effort of cataloguing in the individual libraries; this in turn has resulted in an increased emphasis on the ability to search centralised cataloguing "data bass", usually via terminals linked to computers.Other function performed by cataloguing systems include

- Creation of worksheets for cataloguing, or for input of cataloguing data;
- Prompting of system operators, typists or cataloguers by displaying possible input "tags" nothing misspellings and illogical combinations of data, and leading operators from one step to the next;
- Maintenance of name and subject authority files, i.e., lists of terms used in the system, the sources of the terms, and the cross-references made to and from the terms;
- Production of catalogue cross-reference cards;
- Provision for editing of data for corrections and revisions;
- Arrangement of entries in filling sequence, both for catalogue cards and book catalogues;
- Controlling the movement or location of material, during cataloguing or physical preparation, or while in a cataloguing arrearage;
- Ordering of catalogue cards from a central source;
- Selective listing of catalogue records by specified criteria, such as subject, geographical area covered, data of publication, or combinations of such criteria.

The mechanical means used to perform these functions have generally followed the technological developments described, but with some important differences and additions. The following types are identifiable:

- Systems based on unit record equipment;
- Systems based on automatic typewriters and associated equipment driven by punches tape, magnetic tape, or punched cards;
- Systems using card-actuated sequential cameras;
- Off-line batch-processing systems using computers;

— On-line systems using various types of terminals for searching for initiating catalogue card or book catalogue production (with the production itself usually done off-line); and sometimes for reference by users to the catalogue data.

The first three categories, as well as off-line systems implemented before the development of the MARC (Machine-Readable cataloguing) Project at the Library of congress.

Development of Cataloguing Applications

The earliest cataloguing applications used it record equipment to produce catalogues in book form. Typically, the catalogue information was keypunched, and the resulting punched cards were then sorted into filing sequence (either manually or with the assistance of mechanical sorters) and passed through a tabulator to produce printed pages. In many cases, these printed pages were photographed and then reproduced by offset printing to provide a large number of copies of the catalogue. The method is simple, and once created the machine record in be roused many times. There are marked disadvantages to this method, however:

— Large files of punched cards must be kept in sequence and handled manually every time a catalogue is reduced, or when additional records are sorted into the file;

— The typography available is primitive, without lowercase letters, diacritics, or special characters;

— Only a limited amount of data can be placed on each card, so either extremely brief entries must be used, or multiple cards must be used for each entry, with the cards tied together by a special number created for each entry and punched in each card;

— Only the simplest filing arrangements can be handled mechanically; the rest must either be done by hand, or simplified filing rules must be adopted.

Despite these limitations, a large umber of book catalogues have been produced in this manner. The Library of Congress began experimenting with the process in 1946, and the catalogue of Serial Titles Newly Received (now called New Serial Titles), which the library began publishing in 1950, was probably the first successful, nonexperimental punched-card catalogue. Only a year later, in 1951, the King Country (Washington) Public Library system began a book catalogue system with separate catalogues for each individual library, supplemented by weekly lists.

The punched cards were reproduced on an IBM 513 Reproducer for as many added entries as needed, arranged in the desired sequence, then run though an IBM 402 tabulator to produce the pages. Only one punched card was used for each book producing what later came to be called a "short-title" catalogue or "book list". Similar catalogues were produced by the University of Rochester, the New York State Library, and a number of special libraries.

In most cases, the book form was not intended as a replacement for the library's traditional more available to widely scattered patrons. At Rochester, for example, the "primary purpose" was to put an accurate list of all books and journals in the science libraries on the desk of each member of the faculty and research staff in the departments served by each library. Not every library keeps its card catalogue, however; since the 1960s, the Lake Country (Indiana) public Library has relied entirely on a brief-entry book catalogue produced on an IBM 403 with 15 part paper. The catalogue is in three sections author, title, and subject— which simplifies sorting and consultation. A closed-circuit teletype system connecting the 14 branches and their Technical Processing Center is used to provide additional information, as well as for other purposes.

Perhaps the earliest example of a machine-produced book catalogue with reasonably full entries was the Los Angeles Country Public Library catalogue produced in 1955. At that time, there were 114 branches and services outlets, only 25 of which had card catalogues, and even these were incomplete and softly to maintain. Meeting with the branch librarians resulted in decisions to include the full author, full title, edition (other than the first), data of publication, volumes if more than one, and a short annotation where appropriate for the subject catalogues were issued in eight parts: for adults, there were separate author, title, subject, fiction, and foreign language catalogues; and for children there were separate author, title, and subject catalogues.

The subject catalogues were the most complete, as they included annotations as well as "see" and "see also" references, and analytic for anthologies. The author catalogues contained most of the bibliographic data except for annotations, and the title catalogues were even briefer—a straight alphabetic listing, with only the authors and class numbers given in addition to the titles.

A number of other public libraries adopted the Los Angeles Country system or one similar to it in the early 1960s, including the Montgomery Country (Mary land) Public Library, which (in a noteworthy example of library cooperation) was able to obtain from Los Angeles Country a full set of the latter's book catalogues, the subject heading code book, a detailed description of keypunching procedures, 220,000 subject heading cards, and even the wired control panels for the IBM 407 tabulator to be used! Studies indicated that the cost of the Montgomery Country catalogue was about the same as its card catalogues, with the added advantages of multiple copies for the branches (and incidentally for the country's secondary and elementary schools) and centralised control of error correction and filling. A few unit record cataloguing systems produced catalogue cards instead of book catalogues.

Automatic typewriter systems: Automatic typewriter systems were used a almost exclusively for production of catalogue cards. In such systems, keying the data does not produce punched cards: instead, a spool of either punched paper tape or magnetic tape is produced, and printing is done not on a separate machine like the tabulator, but on the same machine that does the input—that is, on an automatic typewriter. The primary advantage of such systems is the improved typography: lower case as well as upper case is normally available, diacritical marks can be used, and some such systems subject headings in red, mimicking manually typed cards to such an extent that they are indistinguishable.

Other advantages are the elimination of the need for great care in keeping sets of punched cards together, and the ability of most libraries to justify having the equipment housed in the library for its exclusive use, rather than having to depend on a centralised data processing facility for machine time and assistance. Most of these early systems used Flexowriters (paper tape machines) and auxiliary devices that made it possible to program most of the operations involved in producing sets of catalogue cards. The U.S. Naval Postgraduate School in Monterey, California began using such a systems in the mid-1950s and the California State Library in 1958.

In the latter system, the cataloguing data were type on the Flexowriters, which simultaneously prepared two tapes, one for the unit card data and a second one containing the added entry headings. The unit card tape was spliced it no a loop and placed on an auxiliary paper tap reader, the heading type was placed on a second reader. The unit card tape to reading the heading tape at the proper points in the operation. The heading tape contained similar codes for carriage returns and indentations, to that the headings as well as the rest of the data were located properly on continues catalogue card stock.

Later, the continuous stock was separated into individual cards. The University of Missouri began a more extensive Flexowriter system in 1963, and developed a special cutter to separate individual cards from the continuous stock, thereby eliminating the objectionable serrated edges associated with pre-perforated stock. At one time as many as 250,000 cards per year were being produced with the system at an average cost of six cents per card. A similar system, based on Missouri's was developed at Washington University and began operation in 1964. All operation codes were inserted automatically by the machine itself from 'program' tapes, so that the operators were relived of the need to memorise codes and detailed procedures. Programs were also developed which automatically adjusted the spacing of the data for exceptional circumstances; if an added entry heading required three or more lines, for example, the rest of the catalogue card information was automatically lowered the required number of spaces.

As the Flexowriters typed the main entry cards during the automatic production phase, they also produced yet another tape, which was used to create several by-products of the system. The tape was accumulated on a self-winding reel until cards, in punched-card format, for use in an automated circulation system. Next, the tape accompanied the book truck to the processing room, where it was placed on a specially built Flexowriter, equipped with a Se-Lin labeling device, which automatically produced the labels for the books.

The by-product tape could also be passed through the 047 to produce decks of punched cards for processing by computer to produce accession lists and specialised bibliographies, although in actual practice this capacity was never used except for demonstration purposes. Other Flexowriter systems were installed at the University of Colorado, the University of British Columbia, Cleveland State University, and Georgia Institute of Technology, latter two based on the Washington University and Missouri systems.

Another brand of automatic typewriter—the Dura Machine 10—formed the basis for similar systems at Boston University and UCLA. Virtually all of the automatic typewriter systems were later abandoned, how-ever, because of constant difficulties in keeping the machines running properly and because of their relatively slow speeds compared to computer-driven printers. A closely related group of machines called the IBM 870 Document Writing System was used by a number of libraries in the early 1960s to produce catalogue cards.

The machinery was operated by punched cards instead of paper tape, but in other respects similar to Flexowriter systems. The Los Angeles City Schools began using the 870 system in 1962, with the initial goal of reducing serious back-logs in catalogue card production. The backlogs were indeed eliminated, but for this type of application. The backlogs were indeed eliminated, but for this type of application the machines had many of the same drawbacks as tabulating equipment. Keeping sets of punched cards together as tabulating equipment.

The librarian in charge saw this not only as an operational difficulty but also as an educational problem: "Since the cards are produced for teaching-learning situations as well as for finding purposes," she pointed out, "this is a serious drawback, especially since there is an increase in the study of foreign languages in the schools." The cost per set of catalogue cards for the new system was about the same as for the previous manual system.

Other 870 systems were implemented the following year, at the Anaheim (California) Public Library Yale University's Medical School, but by the mid-1960s the 870 was largely superseded by IBM's new 1050 system a similar set of equipment that could use either punched cards or paper tape (Or both), and could be operated independently or connected to a computer. It also used a ball-shaped printing mechanism like that of the IBM Selectric typewriters which permitted the substitution of interchangeable printing elements, and thus different type fonts; a special typing elements, and thus designed for library use. Early library users of the 1050 equipment were the Washington University School of Medicine and the University of Chicago; both of these later abandoned the equipment, however, because of frequent mechanical difficulties. By the late 1960s, the 1050 system was itself superseded by another new IBM system, the MTST (for Magnetic Tape Selectric Type-Writer).

As the name implies, it is also based on IBM's Selectric typewriter, and operates by means of magnetic tape rather that either punched cards or paper tape. The equipment has been widely used for the production of catalogue cards, using procedures similar to the earlier Flexowriter systems described above. The University of California at Irvine installed such a system in 1968, Cornell University in 1969, and the University of Houston in 1970. At Cornell and Houston, however, the MTSTs have since been replaced because of their cost, and Houston has installed a set of similar automatic typewriters called CPTs at about two thirds the cost of MTSTs. Tape oriented automatic typewriters are becoming increasingly popular for business office use, and several models competitive with the MTSTs have recently become available.

On unique system of historical interest should also be mentioned here. In 1961 the Itek Corporation developed a system for the Air Force Cambridge Research Library in Bedford, Massachusetts, which used flexowriters for input of the catalogue data and for printing of the catalogue cards, but not for programming or arrangement of the cataloguing information.

Instead, a special prototype machine called the "crossfiler" was used to expand and arrange the data after input and before printing it out on the Flexowriters. Input techniques were based on what its developers called a "machine-interpretable natural format," that is, the natural typing operations (carriage returns, tabulating shifts, and spaces) were used in a fixed sequence to format the data during processing as well. The technique was simple and effective, reducing greatly the amount of coding that had to be added to the record and in this respect presaging the development of "format recognition" by the MARC project at the Library of Congress some years later.

Sequential card camera systems: While the development of catalogue card systems was in a phase dominated by automatic typewriters, book catalogue systems passed from the unit record era to the use of another type of specialised equipment, the card-actuated sequential camera, some examples of which are still in use at this writing. There are several varieties, but in all such systems, the cataloguing data is typed directly onto cards, and lower-case chains and computer-driven photocomposition devices has lessened this problem.

One of the earliest computerised catalogue card systems was developed in the early 1960s as a by-product of an information retrieval system called SHARP (Ships Analysis and Retrieval Project), a joint venture of the Navy's Bureau of Ship Technical Library and the Devid Taylor model Basin Applied Mathematics Laboratory.

In addition to subject searches and information retrieval, various printed products were generated, among them catalogue cards for all documents entered in the system by 1963, IBM had also begun to produce catalogue cards using computers. From the standpoint of its influence on later developments, however, the most influential of the early systems was without doubt the developed at Yale University. The Yale Medical

Library had begun producing cards on an IBM 870 system as early as 1963.

By 1964, the library was using an IBM 1401 computer to "expand" the catalogue data and create a full set of punched cards, with a "decklet" of cards representing all the information of each catalogue card in the set. These punched cards were then used to drive the 870 system described earlier. Later the same year, an IBM 709 computer was also incorporated into the system to sort the data into filling order before the cards were punched. By 1965, the library and abandoned the 870, and was using the computer printer with a special chain containing an expanded character set to print the cards.

This system was used until 1971, when it was finally discontinued, for two reasons: maintaining the growing data base on magnetic tapes (as opposed to disks) had become increasingly cumbersome, and the computer center had brought in an IBM 360 computer to replace the 7094 for which the system was designed. By this time, personnel at Yale had come to feel that systems designed for one library along were rapidly becoming obsolete, and that the expense of reprogramming the system for the 360 could not therefore be justified.

Off-line book catalogue systems: Although the general procedures are the same, computerised book catalogue systems differ from card production systems in several important ways: storage of the bibliographic data between issues of the book catalogue is normally required; entries input between editing must be interfiled; and arrangement of individual entries in the desired filing sequence for book catalogs is complex, whereas most card systems arrange, if at all, only a relatively small number of cards in rough filing sequence before printing them out.

Early book catalogue systems were characterised by the use of elaborate input coding forms, ruled into 80 columns like an IBM card. Catalogers or their assistants were required to print each letter carefully in a box, and to learn and assign a large number of codes identifying particular data elements. Later systems used worksheets with the codes (or "tags" as they same to be called) already pre-printed on the forms, and still later systems employed techniques which allowed conventional catalogue cards to be "edited," that is, to have the tags merely added to data already printed on catalogue cards or slips.

Processing of the catalogue entries after input normally involves not only duplication of the entries under each heading, but also a significant amount of computer editing (nothing missing data or illogical coding, for example) and several proofreading and error correction cycles. The computer must also sort entries into multiple sequences, since most book catalogs are divided catalog—that is, instead of being arranged with all entries in a single filling sequence, the entries are divided into an author catalog, a title catalog, and a subject catalog, or some similar combination.

When all of the data have been edited, proofread, corrected "expanded", sorted and arrange in the proper sequence, the final step-production—can begin. This may be on a conventional computer line printer (sometimes with a special uppercase and lower-case chain); with a computer-driven photocomposition device; or on a computer output microfilm (COM) device. The earliest project to use computer line printers to produce full-scale book catalogue was also one of the most successful.

In 1963, the University of Toronto was asked by the Ontario government to develop basic collections of approximately 35,000 volumes each, complete with the necessary catalogs, for five Ontario universities. In the ONULP (Ontario New Universities Library Project) that resulted, complete bibliographic information for the volumes purchased (including even the thickness of each volume) was keypunched; stored on magnetic tape by an IBM 1401 computer; edited, sorted and arranged on an IBM 7094 computer; and printed out in book catalogue form on the 1401. The computer printer was equipped with a specially designed chain which contained uppercase and lowercase letters and a large number of diacritical marks (100 character in all).

The computer print-out was reduced to 60 per cent of the original before printing plates were produced so that more information would be contained on each page. Programs were written in COBOL (Common Business Oriented Language), so that the five new institutions could modify them more easily at a later data if they wished, and to make the initial programming task easier and more convenient, but the resulting programming task easier and more convenient, but the resulting programs were unusually lengthy in terms of processing time, a common problem with COBOL programs. Another early computerised book catalogue project was especially significant in terms of the formation and programming techniques employed.

The Boeing "SLIP" system was one of the first to provide a generalised set of programs which could produce a variety of printed products with only minor change in the specifications in -cooperated inn the input to the system, and it was also one of the earliest to use so- called "table-driven" programs for library automation purpose. A total of twelve different tables were constructed by the system's programs in the computer memory, and used to sort input data into varies categories, depending both on the nature of the data (whether it was author, title, subject or imprint data, for example) and the eventual use to be made of it.

At the production stage, the programs then used the tables to rearrange the data by the broad group specified for the final product, and to arrange data within these groupings by filing sequence. The book catalogue produced by the system was divided into an author catalogue and a title-subject catalog, the latter containing both conventional subject headings and "alternate title entries... generated by rotation of the complete titles about the normal indexing locus, the left margin these so-called "permuted" titles, which placed various significant words in the titles at the filing piston and thus used the titles themselves to construct a different sort of subject approach, were to become common features of later systems.

A modified form of the Library of Congress filing rules was used for the book catalog. Titles and permuted titles were sorted full length, continuing on to the first six characters of the author's name if necessary; similarly authors were sorted on their full length, continuing into the first six characters of the title if necessary. Friden Flexowriters were used for input, because the typewritten copy created at the same time made proofreading and correction easier, but the system was also designed to accept punched-card input. The paper tape from the Flexowriters was converted to magnetic tape using an IBM 1401 computer, the subsequent processing being done on an IBM 7094 computer as in the Toronto system.

In retrospect, one aspect of the Florida Atlantic system that seems especially curious is that despite its elaborate coding forms and input procedures, no proofing or editing procedures were established and no correction program was written. Not surprisingly, the first edition of the catalogue contained a remarkable number of errors; blind cross-references abounded, and William Shakespeare "appeared in at least four different places in the subject catalogue and at least four in the author catalogue due to variant spellings." More serious was the fact that although its automaton efforts were well publicised before implementation, no a attempt was made to inform the rest of the library community of its subsequent experiences.

By contrast, the difficulties encountered by the Washington University School of Medicine Library its book catalogue project have been frankly and fully publicised. The first WUSM catalogue was issued in 1965, and contained only entries cataloged between January and September of that year. The computer print-out used for production of the pages was not photographically reduced in size before the printing masters were made, so even for the relatively small number.

The staff, on the other hand, found dealing with the new system a serious difficulty. Part of the problem apparently arose from the fact that the system had been designed with the assumption that bibliographic information generated at the acquisitions step could be used for the. Catalogue step without rekeyboarding, but in the experience of this library idea did not work out. Series might be ordered under a series title, and cataloguing personnel were then faced with the necessity of expanding the single record into the necessary number of individual ones; on the other hand, acquisitions personnel might create separate records for volumes ordered separately, but considered by cataloguing as part of a set.

Another problem noted was the difficulty of changing printed records; once the catalogue had been produced, the records were there until the catalogue had been reprinted or superseded. The Nursing Schools Library was disbanded, but the records for it in the catalogue could not be readily deleted, so "it became necessary to paste a general statement on the cover of each that updating." 44 Problems also were attributed to the small size of the computer memory used; the Dental Library holding were to be added, for example, but only by inputting them little could this be accomplished. After four months or struggling, the library abandoned the book catalogue system entirely. Fortunately (and wisely in this case), the production of catalogue cards had been continued all along.

By the mid-1960s, computer-produced book catalogs also became available from commercial firms. One of the first libraries to obtain a book catalogue in this way was the Baltimore

Country Public Library, which contracted with Documentation, Inc., of Bethesda Mary land for one containing the 50,000 entries cataloged by the library since 1959. Input was done on Flexowriters, processing was done on the firm's IBM 1401 and 1410 computers, and the printing was done with the same "expanded character set" printing chain mentioned earlier. The first catalogue indicated which work were held by which branches, but these location indicators were later eliminated.

The time consumed and the cost involved in updating them were out of all proportion to their-use (and) in addition they were never accurate. If a book is not in one agency it can be reserved through intra-library loan and under ideal conditions, it can be in the hands of the borrower with two days. If a borrower has to have the book immediately he can call intra-library loan and get the names of the branches owning it. He can then call the branch and have it held for him to pick up.

Some 250 copies of the catalogue were produced, over half of them for the country's 125 public schools. By 1971, at least two other public library systems in the area using the BCPL data base for titles also in their libraries, thus avoiding the costs of converting these entries to machine-readable form. The St. Louis Junior College District reported in 1965 that it had contracted with the Alanar division of Bro-Dart Industries sot produce a similar catalogue with brief entries.

A code number was added to each entry, referring the user to fuller information on the same titles in a photographic reproduction of Library estimated that approximately $9, 000 per years was saved in salaries formerly required for clerical filing or cards, plus about $2, 250 per year through the elimination of catalogue card reproduction. Although book catalogs were (and have continued to be) of interest primarily to public libraries, Toronto and Florida Atlantic were not the only universities to intimate book catalogue projects during this period.

In 1964, Stanford University had commissioned one of the landmark studies in this field, 50 comparing the costs of various methods of book catalogue with the costs of a card catalog. 51 Based on these studies, a project was begun to produce a book catalogue for the Meyer Memorial Undergraduate Library, which was scheduled to open in 1966. The first edition was produced in time for the opening, and subsequent editions have been produced annually, supplemented on a quarterly basis. The first editing contained entries for 25,000 titles and by the fourth edition in 1969 the total had reached 60,000 a limit of 100,000 is planned.

Input is by punched cards, with an IBM 7090 used for the sorting and a 1401 to print, again employing the special printing chain mentioned earlier. A cost analysis based on the 41,000-title second edition issued in 1967 indicated total costs of $38,670, some percent more than original estimates but very close to the estimates made during the system design stage.

As an alternative to printing the final page copy on a computer printer, a computer-driven photocomposition device may also be used, Photocomposer of this type read information from tape and by projecting an electron beam on the surface of a cathode ray tube produce pictures of each character in turn, much like the images on a television screen. These images are copied on photosensitive paper at a very high rate of speed (several thousand characters per second) and at a very high resolution (hundreds of lines per inch). The legibility of the characters, combined with the fact that the resulting pages are much more attractive and readable that those produced by computer line-printer. The costs of this method prevented widespread use in the earlier days of book catalogues, but the cost per page has now been reduced to the point that the number of catalogs produced in this fashion is rapidly increasing.

ON-LINE CATALOGUING SYSTEMS

The earliest on-line cataloguing systems used routinely for production of catalogue cards began in 1970 at the Shawnee Mission Public School system in suburban Kansas City. IBM 2740 and 2741 type-writer terminals were used to input bibliographic data, and an IBM 360 Model 40 computer to do the processing and print catalogue cards. A generalized package of programs called FASTER was obtained from IBM, thus avoiding much of the programming effort (and expense) such a system normally requires. Unit costs to create and key data, print cards and labels, and prepare items for delivery were estimate at $2. 50 per item, based on an estimated 100,000 items to be processed in 1971.

In 1972 the cost of cards and labels was calculated at $2.25 per item, or $2.53 if the various listings, reports and other printed products were included. The development of the system write that " the librarians are satisfied that direct access to master library disk files is an efficient, accurate, and economical method of creating and updating bibliographic can holdings information.The public schools of Oregon were also early user of an on-line system using the same terminals.

A large, generalized computer system located in Eugene was already available. This systems, called OTIS (for Oregon Total Information System), utilized an IBM 360 Model 50 computer with 77 keyboard terminals located in various buildings around the state. Three of these terminals were placed in the Springfield School District's Library Processing Center, and approximately 43, 000 titles, representing the complete holding of the district's libraries, were then entered and edited on-line. By 1971 catalogue cards were being supplied for several other libraries.University systems also began to be implemented in 1971.

At Northwestern University, 2470s were connected to a small IBM 370 Model 135 computer in the Administrative Data Processing Center. The system as implemented is a "total " system in what it provides information or titles in process as well as those cataloged, plus bibliographic and holdings information for approximately 35,000 serial titles.The machine-readable files were originally on a data cell (a large and relatively inexpensive form of computer storage), but when this became operationally inconvenient the records were transferred to portable disk packs. Lack of tape drives in the Data Processing Center created another inconvenience, necessitating; tape drives were later added for this reason.

The most serious equipment limitation, however, was the "display" rate of the typewriter terminals which is only 14 characters per second, "far from ideal when long bibliographic records are involved." This difficulty in communication has been the chief reason for the use in later on-line systems of cathode-ray tube terminals with their much faster and often almost instantaneous response. Georgia Tech, whose work on other methods of catalogue card and book catalogue production was noted in earlier chapters, began inputting data in 1971 with Model 33ASR teletype terminals on line to their Univac 1108, and the University of Toronto began the same Year, using Model 1 KSR37 teletype terminals on line to a Xerox Sigma 7 computer. The toronto teletypes were modified to include a large number of characters—seven more, in fact, than the MARC character set.

The toronto system also uses a Xerox Graphic Printer which provides upper and lower case type fonts and various type sizes, thus allowing more characters to be placed on a single card and reducing the number of second (continuation) cards needed. Several hundred thousand records have been input at Toronto using the on-line system, and another 700,000 by a separate keypunch activity. Commendably, the university offers the services of the system to other libraries in the area.

The largest best know and most influential cataloguing system the Ohio College Library Center in Columbus, Ohio, know universally as Ohio College Library Center (OCLC) — also began its on-line operations in 1971. Perhaps no other system has had as much effect on library practice in this country; indeed, if the development of MARC is the most significant accomplishment to date library automation the development and operation of the Ohio College Library Center is probably the second most significant. This is not primarily because of the technological expertise employed in the development of the system (although that is great) but because it is the largest and most efficient shared cataloguing system yet implemented.

OCLC produces catalogue cards economically, as do many of the other systems already described, but its important lies in the fact that it provides a common cataloguing data base for searching and consolation by almost 350 libraries. The margin of savings achievable by any system of catalogue card production is small in relation to the total budget of most libraries, but the cost of performing the intellectual task of original cataloguing is much larger; the OCLC system provides significant savings because both MARC records and the cataloguing records prepared by all member libraries can be displayed of the terminal screens in any one member's library, thus obviating the need for any library but the first to catalogue it.

Catalogue cards can then be ordered from Columbus and arrive in a few days, formatted to a member library's specifications and arranged in rough filing sequence.The Center was founded by the Ohio College Association in 1967 with over fifty members "to operate a computerized, regional library center to serve the academic libraries of Ohio". From the beginning, it was agreed that records compatible with the MARC II format would be used in the Center's cataloguing system so that the Center could play a part in any future national systems.

By 1970, the first OCLC operation, an off-line card production system was functioning. Members submitted IBM punched cards containing the LC card umbers of books for which they wished catalogue cards, along with their own call numbers if they were unwilling to accept LC's and OCLC then ran these requests against the MARC II data base. If the titles were in MARC, catalogue cards were then produced on computers at Ohio State University using programs modified from the Yale project described earlier. Two technical developments associated with this effort should be noted in particular.

Visual display terminals were abundantly available by 1971, but most of them were unsatisfactory for library use; typically only uppercase characters could be displayed, and the number of characters that could be displayed at any one time on the screen was far too low. After beginning with Spiras LTE terminals the Center contracted for production of a visual display terminals specifically designed for the OCLC application, manufactured by Beehive Medical Electronics, Inc.. Called the "OCLC 100" the terminal has been enthusiastically received in those libraries employing it.

The second development of interest is the designs of an efficient searching mechanism. OCLC files could be accessed from the beginning by a Library of Congress card umber, but a method was also needed to allow searching by authors and titles. After some interesting experiments at OCLC and elsewhere, a search "key" technique was devised which requires the terminal operator to input only six characters: either the first three characters of the author and the first three letters of the first word of the title not an article and the first letter of the next three words. This technique and the programs written to accommodate it reportedly retrieve five or fewer titles 99 percent of the time or 80 percent of the member libraries; additional standardized procedures reduce the number of replies in the remaining one percent to less than eighteen with very few exceptions.

Approximately half of the tilts were from MARC records, the other half from participating libraries. And the number of participating libraries was growing rapidly, following a decision in 1973 to allow other regional network to connect to the system. The Pittsburgh Regional Library Center. For many of these customer as for Ohio members, the economic justification for a participation in the system is more than the saving on cataloguing costs. As the price of library materials has risen, the need to share regional library resources has become increasingly urgent, and the OCLC system provides a convenient and effective means of doing so; precise statistics are difficult to come by, but it is clear that more and more participating libraries are using the OCLC system to find and request a copy of a book owned by another library, rather that to catalogue a copy already owned.

Indeed, the implications of the system for resource sharing have often been felt to be as significant as the implications for savings in cataloguing costs. Most importantly, the library wanted to improve communication with the system through the use of visual display terminals, which would al-low catalogue card information to be displayed on the screen and edited before being printed. One problem that plagued the old system may still be a problem with the new, although the library appears to be taking steps to avoid it. In the old system, the computer used was changed from a 360 Model 30 to a 40, then to a Model 50, then to a Model 65, and somewhat later to the present 370 Model 168, all with inevitable delays.

Significantly the new system is also designed so that eventually it can support other user, whereas the old system supported only the University of Chicago library. Perhaps ironically, the part of the system that makes it accessible to multiple user is called CHASM, 28 symbolizing the wide gap between a single-library system and a cooperative one. Whether

Chicago will bridge this chasm in time to attract users away from such established and growing cooperative projects as OCLC remains an important question.

Perhaps the most sophisticate technical processing system presently in operation is the system at Stanford University. Development of the system began in 1967, but its first portion operations did not occur until 1972, when the capability of searching the MARC data base and using it to produce catalogue cards become operational. A large IBM 360 Model 67 computer is used with a small PDP 11 computer as an intermediary and a number of Sanders 804 programmable display terminals. The searching capabilities of the system, developed for the most part under another program called SPIRES (for Stanford Physics Information Retrieval System) are its most impressive feature with many files and indexes on - line and numerous automatic promptings and suggestions from the computer to the terminal user as a search proceeds.

For cataloguing, the user consults one of eleven terminals located in a single area shared by Acquisitions and Cataloging, and following the instructions displayed progressively on the screen calls up the bibliographic information generate in the acquisition process, makes any necessary changes, and adds copy numbers and shelving locations symbols. The end of this process signals production of cataloguing products, including not only catalogue cards and necessary cross-references sequence, but also two kinds of labels, produced on an IBM 2741 terminal with a Se-Lin labeling device attached.

One label is for the spine, the other for the inside of the book in connection with circulation use. If the record used in acquisitions was not derived from marc, the cataloguing department may also search the on-line BALLOTS/MARC files to see if a MARC record is now available. If so, the user may substitute the MARC record for the preliminary acquisition record and add information as needed. If no MARC record is available, original cataloguing does not necessarily have to be performed at this point, because the system is designed to continue searching for MARC records automatically, at least in categories for which MARC data are expected.

If the system later finds a MARC record, the library is notified automatically; in one is found within a specified period of time (up to eight months), the search is terminated and original cataloguing is then performed. In the event that no MARC record is found either on receipt of the item or after the automatic search period, the terminal presents a pretagged from to facilitate input. Editing of the input may be done directly on the terminal or a "catalog data sheet" which the cataloguing personnel use for making handwritten changes that clerical staff the enter via the terminals.

The catalogue data sheet may appear redundant, but it has the advantages of providing a portable version of the proposed catalogue or other sources of information for checking. In this respect, the BALLOTS system has overcome a disadvantage noted for the OCLC system: the physical problem of checking information retrieved on a visual display terminal against information located elsewhere.

Foundation support for the project has been used solely for development and Stanford has paid the operating costs, so when the grant funds eventually disappear the operation of the systems presumably can continue. Nevertheless, certain operational aspects of the system are expensive, and Stanford shares with Chicago the danger of having a glorious white elephant on its hand unless these costs can be spread over a broader base—presumable by supporting the technical processing activities of other libraries as well.

In Stanford's case, work along this line is well advanced. With a new grant from the Council on Library Resources of almost $350,000 an-nounced in early 1975. Stanford will alter the system's software to develop BALLOTS as a California library automation network, and in-cidentally enable it to handle all of the diacritics in the MARC character set. The California State Library has also arranged with Stanford to make BALLOTS available to seven public libraries as the first step toward a California Public Library Automation Network (PLAN). If these trials are successful, a West Coast net work of the OCLC type may emerge, with interesting and important consequences.

Commercial Cataloguing Systems

Commercial firms have also become interested in the possibilities of on-line systems. The System Development Corporation began development of a project called LISTS (for Library Information System Time-Sharing), which would allow libraries con-nected to the system to perform a variety of technical processing tasks, including not only production of catalogue cards but also printing of orders and acquisition reports, maintaining outstanding order and in-process files, and even circulation control.

The system began in 1968 with seven libraries (two public and four academic, plus the SDC library) connected by ASR33 teletype terminals to a 360 Model 67 located at SDC headquarters in Santa Monica, California, but it has never been seriously marketed. The Information Dynamics Corporation announced a service called BIBNET, which uses Datapoint 2200 or Hazeltine 2000 terminal connected via regular telephone lines (though Tymshare's TymNet service) to a computer owned by System Development Corporation. The Datapoint terminals have not only a keyboard and a cathode-ray tube but also a small computer memory and provision for insertion of two magnetic tape cassettes containing programs or data; in effect, therefore, they are minicomputers themselves.

When a desired record or information is located by searching. It can be stored on one of the cassettes and them called up on the screen at a later time and edited without tying up expensive telephone lines. GE Terminet printer terminals are also available, with rated speeds of either 30 or 120 characters per second. The printers can be used for production of catalogue cards, or printing of proof sheets to be checked against an existing card catalogue. Searching may be done by the LC card number the International Standard Book Number, by author, by title, or by use of a "Ruecking algorithm" (a code made up of several characters from the author and title, as in the OCLC system).

In early 1975, company was still in the process of leading MARC record into the system, but brief "index" entries for many records were already available, providing the full author, titles and imprint, as well as LC card number and ISBN. If the full bibliographic record is desired, a key on the terminal can be depressed to ask personnel at IDC headquarters to search manually for the full record, convert it to machine-readable form, and send it back, several days later. The New Mexico State Library has used the systems and a Hazeltine terminal to add holding codes for eighteen libraries in the state, and then search for locations on request from interlibrary loan librarians around the state. SDC's ORBIT program is used for the searching operation. The costs, in addition to the cost of the terminal and its installation, are $45 per hour, plus 20 cents per location added and 90 cents per location retrieved.

All of the systems described above use on-line input and searching techniques in connection with the production of catalogue cards; one other large on-line system, however, has book catalogs as its final product instead. This is the New York Public Library system, perhaps the best and certainly the biggest book catalogue system, in existence at this time. NYPL is really two libraries in one-a large research library, and public library with eighty six branches—so two book catalogs are produce, each containing records added since February 1927. The catalogs are produced on a schedule of so-called "continuous re-accumulation". This means that in order to locate a particular tiles the user must consult only the cumulation for that portion of the alphabet, plus the cumulative supplement, plus the old card catalogue for entries prior to February 1972.

To reduce printing costs, all added entries are condensed, omitting "title page extension", publisher and bibliographic notes, "the assumption being that the user who is interested in such data will take the trouble to refer to the main entry." Input to the system is off-line but three 2740 terminals are used for inquiry. An IBM 370 Model 145 computer is used for processing, and production of the book catalogue pages in done by an RCA Videocomp photocomposer.

In 1972, the cost of production was about $100, 000 per year for 550 to 600 copies of the branch libraries catalog, and about $90,000 per year for 200 copies of the Research Libraries, catalog, of which about forty are sold. The costs of development for the system have been estimated at anywhere from $2 million to $4 million, and the annual computer bill is around $360,000; of this latter amount however approximately $200,000 is paid of by renting out time to private industry.

The project traces its origins to the work of a Columbia University doctoral candidate who found that at NYPL approximately 20 percent of its nine million cards showed substantial deterioration. This study led to a fuller report, which recommended that the library cease adding to the card catalogue cards for new titles using automated techniques. By 1968, plan had been developed for new card catalogs which would grow for five years and then be converted to book form; by 1969 however, the idea of using catalogue cards had been dropped, and a decision was reached to use book catalogues exclusively.

The largest program in the system and its strong feature is its authority file control; a single machine-readable file includes all subjects, series titles, and names used in the system, and every new record is checked against this file before being entered in the system. This provides valuable assistance to the catalogers and avoids the problem of variant forms of entry, a problem that has plagued other similar systems, notably OCLC. This feature also makes it possible to change outdated terminology for new, readily and efficiently, both in catalogue headings and in the cross-reference structure.

Other products of the system include processing information lists, cataloguing work sheets and two shelflist cards for each book which are sent to the branches or divisions. Several other book catalogs, including one of the Dance Collection of the Research Library of the Performing Arts, have also been published. At least two systems have been designed to produce both catalogue cards and book catalogs fro on-line input. The Western Kentucky University library in Bowling Green originally began its project as a means of reclassifying from Dewey to the Library of Congress classification. Ten IBM ATS (2741) terminals were installed in 1971 on-line to an IBM 360 Model 40 computer and temporary workers were used to enter data.

The computer printed out proofsheets of the data daily, and these were then proofread and edited by catalogers, reinforced as needed by librarians from other parts of the library. Labels, pockets and book cards were produced of reprocessing of some 250,000 items, and both catalogue cards for a divided card catalogue and book catalogs were then printed. Since the reclassification project, catalogue cards and book catalogs continue to be produced on a regular basis.

The UCLA Biomedical Library's cataloguing application differs from most other in that it makes no attempt to use MARC records, and is designed to accommodate only "original" cataloging. The majority of the records acquired by the

Biomedical Library are not available on MARC in any case, since the Library of Congress by policy avoids adding works in clinical medicine or other areas covered by its sister institution the National Library of Medicine.

The cataloger for the Biomedical Library types information on a "cataloging coding sheet" which a clerk then rekeys on a terminal, the redundant keyboarding action apparently necessitated by the objection of the cataloger to using the terminal. The terminal itself is a Delta Data Systems device with a cassette section added; the information is recorded on the cassette, and periodically the computer reads the information from the cassette into disk storage.

A prooflisting is printed periodically, and is used by the cataloger to make editing changes. Corrections are then input by the typist. Catalogue cards are produced once a week on a 360 Model 20, prearranged in filing sequence, and a list of new accessions is produced once a month. The same programs that arrange the cards are also used for the book catalogue, which is produced primarily or regional interlibrary loan purpose. A three years, $40,000 grant provided funds for development of the system, and operating costs are estimated at $1,000 per month.

A few libraries have taken what some regard as the ultimate step and devised systems that use on-line terminals not only for input and searching but also for regular conciliations of the catalogue file, thereby eliminating the use of catalogue cards and book catalogs altogether. Perhaps the first such system was the Experimental Library Management System (ELMS) at the IBM Advanced System development Division Library in Los Gatos, California, parts of which have been in operation since 1965.

Catalogue information is input via IBM 2741 terminals and is consulted by both library staff and patrons using IBM 2260 display terminals. There are four of these—two in the processing area, one at the reference desk, and one in a carrel. Public use appears to be successful but since this is a small, specialized library it is difficult to estimate the problems that might result if the same techniques were used in a large research library. Before the on-line reference function was implemented, the system produced nine separate book catalogs or listings, and apparently this capability is still used. Spine labels and book pocket labels are produced.

On-line reference to machine-readable catalogue files was also the subject of much intensive study during Project Intrex at the Massachusetts Institute of Technology. Intrex was a multimillion dollar research project, which began in 1965 and was finally terminated in 1973. On idea that received much experimental study during the lie of the project was the concept an "augmented catalog", which would be consulted on-line and would contain not only the usual bibliographical information, but such additional data as the "author's purpose", the "level of approach", excerpts from the text, all except the author being what the project staff referred to as "content-indicating" fields.

Main entry and publisher were deemed helpful only 9 percent of the time, and such fields as edition, place of publication, and physical dimension were ignored completely by the users. A variety of terminals was used, some of them designed by the project and some commercially available. Not surprisingly, those terminals with full screen displays were preferred to those that had only a typewriter for communication from the computer to the user.

The other major effort of Project Intrex that is pertinent here was the attempt to store and retrieve, via on-line terminals, the full text of documents. Storage of the text was in microform, scanned and transmitted electrically, on demand, to the same terminals used for searching and other purposes. Permanent copies could also be obtained, either in microform or in full-size hardcopy. A summary of use during the project indicated that

— Over 80 percent of the experimental sessions involved use of the text-access subsystem. Most user said they found rapid access to full text a crucial element in a fully satisfactory system and that the text-access system, as implemented, was more than fast enough for them.

— The system operated reliably; over 90 percent of the commands initiated by the user result in the specified output.

— Although most user seem satisfied by the quality of the image when text access is used for the purpose for which it was originally intended (the preliminary ex-amination of the document), Most user prefer higher resolution and express the desire to eventually obtains hard copy.

— Users in this experiment employed the text-access system primarily to judge the relevance of documents after preliminary judgments have been made on the basis of the catalogue data. Users employed text access only to a small extent to read document text for information contained...

— Relevance judgments with the full-text system are made primarily on the basis of document text, as such, rather than associated parts such as illustrations, or the abstract. If we assume that academic level (by year of study) is a reasonable measure of a student's depth of knowledge about the subject of his search, the this depth of knowledge is negatively correlated with the utility of full text for the relevance judging function. The more a user knows about a subject, the more he is willing to rely solely on catalogue information.

— The average user spends somewhat more time looking at catalogue data than he does looking at text, but since he looks at catalogue data on more document.

Although the Intent Project has been formally discontinued, it seems likely that the results of its extensive experimentation, especially with the augmented catalogue and full-text access, will be important to designers of similar systems in future years.

4

Library Automation

Computer application in library and information field has made phenomenal progress in industrialised countries where hardware, software and communications facilities are well developed. In view of their technological advancement, they are able to computerise their entire library and information system in the country, with great success. Besides improving services and operations for a better performance, libraries are able to evolve effective computer networks, towards optimum utilisation of resources and facilities. International programmes like UNISIST/ UNESCO, Universal Availability of Publications (UAP) and Universal Access to Information (UAI) of IFLA and a few others have helped in promotion and coordination of information services through out the world. IDRC, Canada, EEC, Brussels and similar other have helped developing countries to strengthen their information infrastructure.

Global information systems, characterised by decentralised input, centralised processing and decentralised distribution, such as INIS, AGRIS, and DEVSIS have been evolved towards universal bibliographic control and dissemination of literature.

Major producers of secondary information services have become international information/database centres for their respective subjects/disciplines. International information network/centres such as Lockheed/DIALOG, SDC/Orbit and ESA/IRS hold 100 or more databases in several disciplines and offer online services to users through out the world. Communication networks like TYMNET, TELENET and ESANET offer communication facilities to international information centres that offer database services. Towards promoting cooperative endeavors, regional networks like ASTINFO, RINSCA and TECHNONET and specialised networks like AIBA, INNERTAP, SEAMIC and others have come about. These are all useful international efforts in transborder flow of information and in diffusion of knowledge through out the word.

The library networks in USA are in a most advanced level of development. The favourable factors responsible for successful development of library networking in USA are long tradition of cooperation among libraries, introduction of library automation as early as from 1960's, advances in information science during 1970's and 1980's and introduction of MARC format by the Library of Congress in 1968. The experiences of USA in library networking are valuable to our efforts in designing the INFLIBNET.Three foremost bibliographic utilities and networks in USA are the Online Computer Library Centre, Dublin, Ohio (OCLC), Research Libraries and Information Network, Stanford, California (RLIN) and Washington Library Network, Olympia, Washington (WLN). In addition, there are networks of different types, levels, etc.

The library networking is based on three major areas of technology, viz., creation of bibliographic information in machine readable format for storage and retrieval, hardware/software for database use and services and telecommunication mechanism for transmission of information. The major point of success of library network depends upon availability of machine readable catalogues in the constituent units. A great progress has been made by USA in this regard. The latest trend is to use CD-From technology for storing the catalogue data. The OCLC, established in 1967, and has a staff of 812 persons (as on 1987) is a non-profit organisation and supports resource sharing among more than 6,700 libraries in USA, besides countries in Europe, Saudi Arabia and Australia. Its union catalogue database contains 13 million records; some 30,000 records are added every week. With the steady growth of OCLC, there has been a progressive decline in the amount of original cataloguing to be done by the participating libraries.

The system hit rate has increased from 66 per cent in 1971 to 94 per cent in 1983. The details about telecommunication system, input sources. RLIN, established in 1978 and which has as staff of 81 persons, supports the cooperative programmes of Research Libraries Group, comprising 36 major libraries and other research institutions. It maintains 6 databases online. The catalogue database holds more than 20 million records.WLN, established in 1972 and which has a staff of 51 persons, provides, more than 250 libraries in seven states of USA, online computerised services to promote resource sharing and automated library functions. The bibliographic file contains more than 3.5 million catalogue records.

Library Network

An information network is a set of interrelated information systems associated with communication facilities, which are cooperating through more or less formal agreements and institutional agreements, in order to jointly implement information handling operation, with a view to pooling their resources and to offer better services to the user. They generally follow identical or compatible rules and procedures. A library network consists of six major components:

- — Information resources
- — Reader or users of information
- — Schemes for intellectual organisation of data/documents
- — Methods for the delivery of information to user- the output
- — Formal organisation—cooperation, whether voluntary or obligatory, in information exchange and utilisation
- — Bidirectional communication links

Based on these components, the essential requirements of INFLIBNET can be identified as:

- — Libraries (University, R&D Institutions, etc.)
- — Computer facilities (Main Frame, Mini & Micro computers
- — Machine-readable databases (Library holdings, Union catalogues, National & international database)
- — Reprographic facilities and
- — Communication link (Telephonic, postal or satellite communication channel).

Application of Automation in Libraries

Compute-based Accounting

The University Library in Newcastle-upon-Tyne has been operating a computer-aided ordering system. The introduction of this system was facilitated in that it makes use of a number of programs already in existence as part of the Newcastle file-handling system. This is a suite of programs written for the English Electric KDF9 computer to facilitate the handling of structured data of the kind found in bibliographical applications.

It is interesting to note that Newcastle University library report a small saving in staff time compared with the previous order system which used edge-notched cards. If computer time were charged to the library, however, the automated system would be considerably more expensive. For new books which are included in the Virginia Kirkus Service pre-punched cards are distributed, one for each title, giving the usual bibliographical information. To order one of these books all that a library has to do is to return the appropriate punched card to the data-processing centre indicating the number of copies needed. This is an interesting example of one way in which cooperation between libraries can make for simpler automated systems, save time, and reduce costs.

At the University of Maryland an acquisitions system is in operation utilising punched-card equipment only. Orders and auxiliary listings of outstanding orders, receipts, and fund reports are made using an IBM 407 accounting machine. "On order" cards are filed manually in the public card catalogue and pre-cataloguing is carried out whenever complete Library of Congress cataloguing data is available. Shaw has recently described an application of data processing techniques to the accounting side of library acquisitions at the University of Hawaii. He shows that accounting costs are reduced to one thirteenth of their previous level by this very simple system which just uses the computer as a calculating machine.

In libraries where funds are limited it is important to remember that modest automation programmers can result in significant savings which can be directed to improving conventional services. A unique computer-aided system for ordering replacement copies of standard books is now operating in Toronto. Public librarians there have selected a standard list of about 2500 titles, all of which are available in paperback editions. Given information on the quality rating of publishers' editions, the computer prepares monthly lists of replacement orders which enable each branch library to have in stock a determined number of copies of all the standard titles, all in good condition.

A number of library systems in the Toronto area are cooperating in this scheme which, because of its size enables substantial discounts to be obtained. A number of other libraries have reported automated acquisitions systems to be in operation. At the library of the University of Michigan a disc memory store is used in the computer processing of order records, while at the Redstone Scientific Information Center a system christened ALPHA (Automatic Literature, Processing, Handling, and Analysis) utilises an IBM 632 electronic typing calculator for key punching the original order records.

Computer-aided Circulation Control System

A typical computer-aided circulation control system is now operational at the University of Southampton. It is being implemented in four stages of increasing sophistication the first of which commenced in operation. The first stage, as a by-product of its operation, produces a book card for each book that is circulated, thus building up an "active" collection of books with books cards ready for automatic input of charging data during the second stage of the system. The final two stages provide for on-line interrogation and updating of the loans file. The price which has to be paid for this gradual method of building up an active collection is the heavy work load involved with the first stage. Borrowers fill in slips for each book they borrow and reserve and data-processing staff keypunch one card for each of these transactions and also one card for each book that is returned.

The advantage of a crash programme is that the circulation system can employ automatic collection of charging data right from the very beginning, and that there will be no need to cater for books being circulated without cards. Although cards will be made for many books which will in fact never be circulated again, it will not be necessary to punch cards for books that are returned as at Southampton.Trueswell has shown that for a university technical library where it is the newest books that circulate most, it is more economical to prepare book cards for books on loan during the loan period than to carry out a crash programme for the entire collection. And if all new acquisitions are supplied with book cards at the technical processing stage, possibly as a by-product of an automated order system, then in time only a small percentage of books will be presented for circulation without book cards. However, in a public lending library where a high percentage of the stock is circulated and books have a long "half-life", a crash programme may be appropriate.

Ruecking has made a mathematical analysis of future circulation growth at Rice University in order to make a more accurate comparison between different methods of circulation control. As circulation increases and as filed grow, linear relationships do not necessarily hold, so that it is necessary to include the cumulative costs of a system when evaluating it. His calculations show that over a 6-year period the total cumulative cost of a computerised system would be about half the cumulative cost of any other system.A number of libraries in the United States have now adopted circulation control systems based on either the IBM 357 or 1030 data-collection units, accepting data from eighty-column punched book cards and readers' plastic badge cards. Among these are Southern Illinois University, Florida Atlantic University, Lehigh University, Oak Ridge National Laboratory, and the University of Missouri.

An unusual system operates at Johns Hopkins University which represents one solution to the slow charging speeds which are associated with computer-aided circulation control. The price paid is an extra operation in the circulation procedure. Charge and discharge data is captured on microfilm by photographing the reader's identification card and the spines of up to twelve or fifteen books being borrowed. After developing the film the data is keypunched and processed by the computer in the normal way. In Germany, two university libraries, Bochum and the Technical University of Berlin, have implemented automated circulation control systems using book cards edge-punched with paper-tape codes for automatic collection of charging data.

Circulation control systems using punched-card machines have not yet been displaced by computer-based systems. One has recently been installed in the Atomic Weapons Research Establishment at Aldermaston. It makes use of the card duplicating facility of an IBM 026 printing card punch to create three transaction cards for each loan from punched book and reader cards.

One card is filed manually in the issues file and the other two are used by a computer to apply various controls to the system; to test for overdue books, for example. Eighty-column cards are filled in by borrowers and then keypunched for processing. A computer is used only for printing letters recalling overdue books. Judging from the experience of the University of Hawaii it would be foolish to assume that computer-based circulation systems are the only ones worth introducing in this modern age. The best system for each individual library can only be arrived at after rigorous systems work and careful costing.

Periodicals listing

Many libraries have produced lists of their periodicals holdings by computer punched-card methods and taken advantage of the opportunities thus afforded for distributing the list widely and updating it easily and frequently. A union list of periodicals held in Institute of Education libraries in Britain was also published in 1966. It was produced with the aid of some programs of the Newcastle file handling system. Experiments are in hand to print new editions of this union list by computer typesetting techniques being developed by the University of Newcastle-upon-Tyne computer typesetting project.

The Atomic Energy Research Establishment at Harwell produced a printed list of its periodicals holdings by punched-card techniques in 1960. The main library at Harwell is responsible for purchasing periodicals for its own collections as well as for those of twenty-two divisional libraries. The use of mechanised methods for compiling the list made it a comparatively simple job to supply each library within the system with a frequently updated union list of periodicals.

The most important British union list of serials produced by mechanical means is, of course, the new BUCOP. Input is from punched cards and processing is done by IBM 1401 computer. In the case of a widely used list such as this it becomes important that computer typesetting techniques should be introduced quickly, even though the present frequently updated list is an immense improvement on the older ones.

An interesting holdings list in the United States is that of the Francis A. Countway Library of Medicine, formed by the merger of the Boston Medical Library and Harvard Medical Library. The incentive to mechanise was the immediate need to merge the records of the two libraries. An impressive number of different outputs have been produced including a KWIC index of current titles.

Semi-automated serials check-in systems have a number of attractive features and have been adopted by several libraries. They can be regarded as an intermediate stage between systems producing a list of holdings and fully automated systems embracing accounting, checking in, and binding routines for periodicals. For libraries which cannot afford to develop a fully automated system directly they are a useful stepping-stone whereby the final goal can be approached by degrees.

In Britain a computer-produced checking-in list has been developed at the City University. Periodical parts expected to arrive during the coming month are listed on a line printer. As the parts are received in the library they are ticked off on this list, making it a simple matter to inform library users about the arrival or non-arrival of a particular journal issue.

This printed checking-in list gives full titles of journals, unlike systems which punch an eighty-column card for each expected part, where title have to be abbreviated. The system is capable of conversion to a fully automated one, which would require only a minimum amount of keypunching to cater for parts which do not behave as expected.

Another semi-automated system has been installed at the Louisville School of Medicine, Louisville, Kentucky. Using only punched-card equipment, check-in cards are prepared for each expected title. These require the manual addition of volume number, part number, and date of receipt to complete the full bibliographical data for each part. A number of libraries in the United States have reported the installation of fully automated computerised systems.

Online catalogues

The very simplest level at which entries for bibliographies and catalogues can be prepared by automated methods is that involving the use of a paper-tape typewriter or the IBM Document Writer. These machines are used widely in libraries nowadays and their routine application is seldom reported in the literature. Two useful articles which explain how they have been used in particular libraries are those by Wilson and Bernstein. Conflicting reports have been published on the relative costs of producing catalogue cards by tape typewriter and by other techniques. Mackenzie shows the tape typewriter to be about 75 per cent as costly as either a stencil duplicator or a small offset litho machine. Moore, on the other hand, finds the reverse to be true-duplicating methods works out to be approximately 75 per cent as costly as tape typewriters. The cause of these divergent results is to be found in the relative amounts of clerical time stated to be required by the two methods.

An interesting variation on these methods has been reported by Fasana et al. Catalogue records are prepared in machine-readable form and in a special machine-interpretable natural format using a tape typewriter. Working from the tape punched by the tape typewriter, a special digital machine-the Itek Crossfiler-prepares a catalogue entries. Running this second tape through a tape typewriter produces the required card set directly. Briefly, this technique obviates the need to use a program tape when added entries are being typed and makes for greater flexibility in the production of added entries of different kinds. There is a large literature dealing with various aspects of the production of card and book catalogues by means of punched cards and computer methods.

Dewey gives a brief historical survey of the earliest catalogues produced by punched-card methods and records that the first non experimental catalogue of books was made in 1951 for the branch libraries of King County (Washington) Public Library System. More widely known are the book catalogues of Los Angeles Country Public Library, first introduced in 1953. Before that date only 26 of the system's 114 service outlets had card catalogues, while the introduction of book catalogue enabled each branch to have a union catalogue of the whole library system. A book catalogue produced by punched-card methods was introduced to the science and engineering libraries of the University of Rochester in 1962. This short-title catalogue, now made using an IBM 7094 computer, is designed as a supplement to the libraries' regular card catalogues and each entry is accordingly limited to one line only. In the latest editions of this catalogue refinements have been made to the sort programs. Abbreviations are filed as if they were spelled out in full and names which have prefixes such as de, le, van, von, etc., separated from the main name, are interfiled with names where the prefixes are not separated.

Among the North American libraries that have made book catalogues by computer are Stanford University Library, where the decision to automate was based on a thorough comparative cost analysis, Florida Atlantic University, where the original intention was that the catalogue should also be searchable by machine and Toronto University Library, which prepared a book catalogue of the combined holdings of five new university libraries in the province of Ontario—the ONULP project.

At Harvard University Library a project is under way to convert the entire shelf list to machine-readable form. In that a shelf list is a classified sequence of bibliographic records, the whole job can conveniently be split into small units and one main class keypunched at a time. The first Harvard class, the Crusades, was completed in May 1965. As with the short title catalogue at the University of Rochester, these printed shelf lists, invaluable as they are, can only be regarded as supplements to the main card catalogue where full bibliographic details are available.

De Gennaro has recently developed this classwise conversion technique into a proposal for creating a national union catalogue in machineable form for the United States. Each class of the Library of Congress shelf list would be converted and machine printout compared with the holdings of other large research libraries. They would add to the machineable records their own locational symbols and titles held only by them, and could print their own book catalogues by selective listing of the master tapes.

The medical libraries of Yale University and Columbia University are co-operating in a project which will eventually provide them with a common pool of books and journals the catalogue of which will be in machineable form and accessible through remote on-line consoles. Kilgour has described programs to produce printed catalogue cards as the first stage in this ambitious project. The programs were written in a symbolic-assembly language to take fullest advantage of the potential of a small computer-a two tape 4K IBM 1401. In another paper Kilgour outlines the advantages to be gained from writing programs in such a machine-oriented language as compared with a higher-level language such as COBOL, which is proposed by Culbertson as a standard language to be used for library data processing.

An unusual computer-produced catalogue on microfilm has been operational in the Technical Information Center of Lockheed Missiles and Space Company since July 1966. Using a machine to compose directly from the computer onto microfilm is stated to reduce costs considerably. Access to the catalogue must be by means of a microfilm reader, but film kept in cartridge makes for simple handling and a Xerox Copyflo can be used to produce hard copy quickly and cheaply if this is wanted.

The Library of Congress project MARC has had a marked influence on libraries' plans for catalogue automation both in North America and Britain. The MARC pilot project, in which cataloguing copy was distributed on an experimental basis to sixteen libraries, began in October 1966 and is still continuing.

The public library systems of the London Boroughs of Barnet, Camden, and Greenwich have had computer-produced book catalogues for nearly 3 years now. Using the services of a computer bureau and accepting severe limitations on the format and layout of entries, these libraries have been able, nevertheless, to improve their public services by supplying all branches with union catalogues, a virtual impossibility with a manual system. Tait and Anderson, of the Department of Librarianship at the University of Strathclyde, Glasgow, have recently described an experiment to test the effectiveness of an ICT package program for library indexing and cataloguing-NIC. They are using citations from Library Science Abstracts as inputs data to the system. It will be interesting to compare the effectiveness of the indexes produced by NIC (Nineteen hundred Indexing and Cataloguing) with the conventional indexes published for Library Science Abstracts.

At the Cambridge Philosophical Library project has commenced to produce by computer a union catalogue of books in the many departmental libraries with in the University of Cambridge. The entire shelf lists of a sample of these libraries are being converted into machineable form and a thorough investigation will be made into the problems of merging the separate catalogues at the University of Rochester. Unless entries referring to identical books really are truly identical the computer will include them both in the updated catalogue.

At the University of Newcastle-upon-Tyne, Hunt is working on a computerised catalogue for the two special collections in the University library. Using programs from the New Castle file-handling system, this catalogue will be indexed by author, binder, provenance, etc. Particularly interesting techniques being developed by the computer typesetting project at the University.

In Germany, book-form catalogues are being produced by computer at the University of Bochum. This is a new university and in fact has no card catalogue at all. Germany has also made an important contribution to computer-aided cataloguing methods by being the first country to produce its national bibliography by computer. Type-setting is by means of a Linoquick machine, a normal line-casting machine which can be controlled by paper tape. The main advantage of this data-processing system became apparent with the first half-yearly cumulation, the Halbjahres-

Verzeichnis. Normally this job takes from 12 to 15 months, but it has now been reduced to 2¼ months.

Cataloguing has three parts :

— descriptive cataloguing,

— subject cataloguing and

— locational information.

The objective of this service is to share by a library the work of cataloguing of publications already done elsewhere in the INFLIBNET. It will enable a librarian to use the catalogue information available in a major, nearby university library or regional centre for cataloging new publications added to his library. In other works, a document is catalogued only once at the time of first input into the network.

Bibliographic Database Services

These services would enable library staff to search bibliographic databases developed/mounted at one or more sectoral and other information centres, in order to disseminate current information and retrieve retrospective information. Anticipatory services like S.D.I. and current awareness service and responsive services like literature search and bibliography compilation by downloading could be offered with the help of the databases.

Bibliographic databases may be considered as of three types:

— retrospective databases in core areas, built on the computer facility at the sectoral and other information centres. The databases comprise bibliographic citations/abstracts of world information on magnetic tapes, which are acquired from foreign database producers as well as locally produced machine readable databases of Indian bibliographic information.

— retrospective databases in peripheral/low usage areas, available through international/low usage areas, available through international search systems like DIALOG, BRS and ESA which can be accessed through appropriate gateways.

— retrospective databases on CD-ROM for secondary information, whenever they are available for useful and trust areas, from which service could be generated nationally.

In science and technology, retrospective databases of core interest which need to be made available in the country are: CA SEARCH for chemistry, BIOSIS for biological science, INSPEC for physics and engineering, COMPENDEX for engineering, MATHSIC for mathematics, GEOREF for earth science, MEDLARS for bio-medicine and AGRIS for agriculture. In social sciences and humanities, Sociological Abstracts for sociology and ERIC for education may be identified. The above areas are of the first order and more could be added later in phases.An integrated information system, besides handling bibliographic information, should also cover nonbibliographic information about ongoing and completed projects, institutions, specialists, etc. The INFLIBNET should make it possible for participating libraries to capture this information from a unified database store at the regional level.

Document supply services

Information dissemination services by themselves are of no avail unless they are backed up by actual provision of copy of source documents. The user is invariably interested in going through the original source document of his interest. For that matter, document supply service assumes a great importance and the INFLIBNET has to play due attention to it.

Inter-library loan

This service enables a library to request another library through the network, for one or more books on inter-library loan basis for meeting the demands of its users. This may also include facility for reserving a book, if it is on lone in the leading library. The actual transmission of the book itself will be through postal mail or other physical delivery mechanism like inter- and intra-city courier services

An efficient and effective mechanism for physical delivery of the document needs to be developed. Success of this service depends upon the cooperative spirit of participating libraries. A code for inter-library loan transaction may have to be evolved and adopted by all the libraries. Libraries having a richer collection will be major providers of this service. Such libraries may need additional support for administering the inter-library loan service satisfactorily.

Document delivery (Fax)

Document delivery (fax) enable a library to request another library for a copy of a document, to be transmitted via telefacsimile. This service may be largely used for transmitting a few pages from documents, say, journal articles. It is to be provided by the Document Resource Centre which are to be identified for this purpose within the INFLIBNET. Problems associated with copyright regulations need to be considered in consultation with concerned organisations like, the Department of Education, Ministry of Human Resources Development. It may, however, be noted that the major users of INFLIBNET belong mainly to research and academic community and that reproduction to a single copy of journal articles for bonafide academic/research pursuits may not meet with any serious problem in regard to copyright regulations. Currently, the paper used for fax reproduction cannot be preserved for more than a year due to the chemicals used in its preparation. It is also expensive. In view of the anticipated high growth in demand for document copy supply service in the INFLIBNET, alternatives for more economical document reproduction methods like image capturing through a video camera to a PC and reproduction using a high speed dot matrix or laser printer should be explored.

Document delivery (Non-Fax)

Document delivery (Non-Fax) refers to supply of hard copy of documents required by users, by exploiting/mobilising the document resources in the country through the Document Resource Centres. Supplying xerox copies of documents is the most convenient method, even though microfilm/microfiche may be preferred occasionally for certain reasons. The Document Resource Centres are to be equipped with one or more heavy duty xerox machines and supporting facilities to handle a large volume of demand for document copies.

The INFLIBNET is a vast network and a major part of the demand for document copies could be met from some library in the network. Still, there will be a residual demand for documents that are not available in the country within and outside the INFLIBNET. There would be a need to procure documents from foreign libraries and documentation centres. Organisations like INSDOC are offering national level document delivery service and these facilities could be availed to meet the demand for documents not available within the country. Alternatively, the British Library Lending Division at Boston, Spa, which is well known for its international document delivery service, can be approached in such cases. The task of procuring documents not available within the country may be centralised at and handled by the Document Resource Centres and the Regional Centres.

Library Collection development

The objective is to help the libraries to identify and select publications which are worth while acquisition. This can be based on usage data available at different levels in the network. After examining the strengths and weaknesses in the document collections in the libraries as well as duplication and based on the specialisation of libraries which are to be designated as Document Resource Centres, the INFLIBNET will gradually evolve a cooperatives acquisition system. The libraries will be enabled to develop their collection on the basis of mutually agreed upon profiles assigned to each of them, so that duplication in acquisition is minimised.

The INFLIBNET's objective of resource sharing implies also sharing responsibility in acquisition, in order that the available total book purchase fund of the INFLIBNET is allowed to go the longest way for acquiring maximum number of documents in the network as a whole.The INFLIBNET will also assist the libraries in the procurement process by providing purchase related information like cost, supplier, ISBN number local agencies, currency conversion, discount rates, etc.

Referral Service

Requests for actual provision of information which cannot be satisfied by service library are referred to another source wherefrom the information can be had. In this service, the users are information can be had. In this service, the users are informed of the sources to the approached to get the required information. The users are put into direct contact with the sources of information.

The libraries will be required to maintain a representative stock of referral directories and tertiary tools for undertaking this service. The referral function will get maximised progressively in the libraries at higher levels of the hierarchy.

The communication system of the INFLIBNET will be used to send referral queries and receive answers.

E-mail

E-mail gives an ability to transfer/receive message among the participants in the network. This is an important facility in the network through which other services like interlibrary loan requests, location search in union catalogues, document delivery requests transmission, referral and reference service and academic communication are implemented. This service will reduce communication delays in switching information among the constituents of the INFLIBNET.

Bulletin board

Bulletin board is a facility to display/view in an electronic bulletin board, news, announcements, etc. with constant updating of information. Several bulletin boards can be made available in the network for each of specific categories of users, node types, staff, disciplines, etc.

Academic Communication

The INFLIBNET can be extended to meet the communication requirements of the academic and scientific community. This may be accomplished through electronic mail, file transfer and bulletin board. Computer conferencing can be also carried out through the network facilities. More sophisticated instruments will be needed by the network to support services like teleconferencing.

Library Automation in India

With the advent of low cost microcomputers, the libraries and information centres are now encouraged to switch over to computerisation. There is also an enthusiasm among the professionals to get trained in library automation. Till to date, nearly 100 library and information centres are known to be computerising their activities. The number is steadily increasing. Some of the major organisations engaged in computerised information handling are:

— Indian National Scientific Documentation Centre (INSDOC)
— Documentation Research and Training Centre (DRTC),
— Indian Institute of Science, (I.I.Sc.)
— Indian Institute of Technology, (IIT), Madras
— Indian Institute of Technology, (IIT), Kanpur
— National Informatics Centre (NIC), New Delhi
— Bhabha Atomic Research Centre (BARC), Bombay
— Tata Institute of Fundamental Research (TIFR), Bombay
— Physical Research Laboratory (PRL), Ahmedabad.
— Indira Gandhi Centre for Atomic Research, Kalpakam, Madras.
— Defence Scientific Information and Documentation Centre (DESIDOC), Delhi.
— NISSAT Sectoral Centres: NICDAP, Lucknow; NICLAI, Madras; NICFOS, Mysore; NICMAP, Bangalore; and NICRYS, Madras.

The UGC National Centre for Science Information (NCSI) at IISC, Bangalore has been undertaking a major effort of offering computerised SDI service from foreign databases, such as INSPEC, CA-SEARCH, BIOSIS, MATHSCI and GEOREF to more than 3,000 user in academic and research institutions.

In agriculture, the ICAR Agricultural Research Information Centre (ARIC) provides computerised information services from AGRIS. The ICMR-NIC Bio-medical Information Centre, New Delhi is engaged in providing online access to MEDLARS and POPLINE databases. The National Medical Library is also engaged in computerisation of its bibliographical services.

Information Centres in R&D units of industrial undertakings such as SAIL, Ranchi; BHEL, Hyderabad; HMT, Bangalore; IDL, Hyderabad; EIL, New Delhi; and MECON, Ranchi have also developed capability for computer based library and information activities. Other notable institutions engaged in computerisation are the Central Secretariat Library, New Delhi; Bureau of Indian Standards, New Delhi; Institute of Armament Technology, Pune; National Aeronautical Laboratory, Bangalore; National Council of Applied Economic Research, New Delhi; NISTADS (CSIR), New Delhi, PID (CSIR), New Delhi, NITIE, Bombay; and VSSC, Trivandrum.

In the university system, the Tamil University at Thanjavur has taken a lead in initiating computer based activities. Thirteen university libraries are reported to be using computer for one or more library functions. The International Crop Research Institute for Semi-Arid Tropics (ICRISAT) near Hyderabad is well known for its progress in computer based activities in information work.

The Department of Electronics, the Department of Telecommunication, etc. have established or have been engaged in establishing data communication networks such as NICNET for Government information by NIC, INDONET for commercial use by CMC, VIKRAM, a packet switched public data network of India by the Department of Telecommunication and SBRTN, a satellite based rural telegraph network for North East India By Space Application Centre/DOT.

There are possibilities of mounting bibliographical databases on these networks for providing online access to users.

An academic and research network, known as ERNET, which could address data communication at various levels—Local Area, Wide Area, Metropolitan Area and Rural Area—is also operating since 1986.

Automation being carried out by libraries is usually in acquisition, serials control, cataloguing, circulation and management statistics. Some progress has been made by the libraries in these areas. A few institutions are cresting and maintaining bibliographic databases in machine readable form and bringing out lists, catalogs, union catalogs, etc. They offer current awareness service.

The foreign, machine readable databases being subscribed/acquired in the country are CA SEARCH BY INSDOC and NCSI/IISc, INSPEC by INSDOC and NCSI/IISC, COMPENDEX by INSDOC, BIOSIS by NCSI/IISc, GEOREF by NCSI/IISc, MATHSCI by NCSI/IISc, INIS by BARC, AGRIS by ARIC/ICAR, INPADOC by NIC, METADEX by EIL and SAIL R&D and Cambridge Crystallographic Data by NICRYS. SDI service is being offered from these databases by the concerned institutions.

India is inputting domestic information to international information systems such as INIS by BARC, for nuclear science, AGRIS by ARIC/ICAR for agricultural science, IFIS by NICFOS/CFTRI for food science, ISORID by DRTC for R&D projects in library and information science, DEVSIS by CENDIT for development science, INFOTERRA by ENVIS for environment and ISDS by INSDOC for serials data. In the process, the institutions are able to obtain global databases from the international information systems.

NISSAT is engaged in promoting the development of sophisticated all purpose library application software packages. However, some important packages now available for use in India libraries are CDS/ISIS (microversion) developed by UNESCO and MINISIS developed by IDRC, Canada.

Private software firms have recently started to develop library application software packages and they are made available on a commercial basis. Some libraries and information centres have developed softwares in-house for specific application, with local expertise.

Serious attention is now being paid to training aspect for developing skills in computer application among the library and information manpower. While, towers long term requirement, training institutions like DRTC and INSDOC place emphasis on information technology in their course curriculum and the library schools in the universities have begun to introduce computer application in the courses being offered by them, towards short term requirement of training the existing personnel, numerous training programmes are being launched by professional associations, institutions and others. NISSAT, UGC, etc. are promoting/sponsoring such training programmes in a big way.

Role of Computers in Future Library Systems

The pattern of automated library service in the future can only be predicted by educated extrapolation of the developments that are taking place in libraries today, bearing in mind other factors which will have an important bearing on libraries. Included among these will be an expected continuing exponential growth in population and the increasing the importance of libraries in the developing countries.

The expansion of knowledge shows no signs of abating, and libraries will also be faced with a continuing proliferation of disciplines, coupled with a trend towards inter-disciplinary study. The volume of publications will continue to grow but there should be an increasing recognition by central governments of the importance of libraries and their functions of bibliographic control and information dissemination. The consequence of this recognition will be greater financial support for library programmes.

The key requirement in such an age is that libraries should be flexible. This will be manifested in all facets of librarianship; library buildings, systems for subject representation, forms of publications and the reader services provided for the library's clientele.

Co-operation between libraries at local, national, and international levels will increase in importance, but it must be flexible so as to ensure that individual, importance, but it must be flexible so as to ensure that individual, seemingly more important goals. Most important of all, perhaps, is the recognition that libraries serve as an intermediary in process of communication between human beings. They are not ends in themselves and the fact that libraries must deal ultimately with people, not merely with inanimate documents and machines, obliges us to design systems which treat both readers and library staff with understanding and respect for their personal feelings and needs.

The computer has a central role to play in such future library systems. As a reliable, competent, accurate, and cheap manipulator or library infor-mation it can bring to fruition many idealistic concepts about flexible library services.

In that it has the capacity to free human beings from much of the repetitive labour that is a necessary concomitant to library work, it can also give us the opportunity to provide personal service to real people. The necessary prerequisite to the wise and proper use of computing machinery in libraries in the future is a body of wise and informed library managers now.

It is possible to distinguish two general directions in which computer aided services will develop. Printed lists prepared by computer processing of a basic corpus of bibliographic records will increase in number and in diversity, and facilities for on-lie interrogation of files held in machine memories will become more widely available.At the present moment computer-äided library systems almost invariably produce as their end product some sort of printed list for human consultation or scanning.

In the future, then, we can expect to see more such lists prepared in circumstances where they can be justified either by reducing costs or increasing services. At present, too, a number of experimental on-line library projects are either operational or under development. It is reasonable to suppose that these will be successful and will eventually find application in certain circumstances.It is possible to divide present-day on-line projects into two broad categories; those involving the automation of one or more library housekeeping functions, and those concerned with information retrieval and the processing of natural language text.

The major area of library housekeeping which librarians tend to regard as being suitable for on-line methods is circulation control. Although no system is operational yet, work is either in progress or under consideration at the University of California, San Diego and Los Angeles, and at Harwell and Belfast. A proposal for on-line accessioning of periodicals has come from the University of Lancaster, and a number of libraries are considering systems for on-line manipulation and cataloguing data.

Assuming that it remains essential to make non-programmable alterations either to ordering data or catalogue data from a central agency before it can be added to a library's own catalogue file, then an on-line visual display console equipped with a keyboard and light pen would appear to be an attractive piece of hardware for doing this. All of these on-line systems, however, involve only minor modifications to existing patterns of library activity.

More far-reaching changes to present-day methods of library use appear to be implied in on-line systems for information retrieval and text processing. Those that come closest to everyday library operations involve the on-line interrogation of a file of bibliographic records. This enables a search to be directed in the first place to the names of authors, editors, etc., titles, class numbers, or coden. Secondary search of a body of records retrieved by one of the above terms can be by means of a subject term; publisher's name, date of publication, place of publication, language, or location.

The details of the system have yet to be worked out, including the bibliographic record format to be used, but the body of records will be indexed in four ways corresponding to the four data categories by which primary access to the file may be obtained. Presumably, search statements can combine more than one term in logical relationship.

A project is also under way to create an on-line union-catalogue for the public libraries of metropolitan Toronto. Here, in addition to the design of the machine system, difficulties have to be overcome in respect of the machine system, difficulties have to be overcome in respect of differences of cataloguing practice between co-operating libraries. It is likely that this project will provide for only limited degrees of search complexity, and will handle only abbreviated bibliographic records in order to minimise the size of file required.Kessler is conducting an interesting experimental project at Massachusetts Institute of Technology which permits on-line searching of bibliographical records by various novel search strategies.

Articles from twenty-one specially chosen physics journals are "indexed" for the system, but conventional subject indexing is not employed except in so far as it is made available in Physics Abstracts. Search strategies can be defined on the basis of journal titles, author's names and—words in titles. Citation-orientated searching is also possible, for example, one can search for all papers which cite a chosen article or alternatively, all papers that have a stated number of references in common with the chosen article. Kessler has introduced the concept of bibliographic coupling to describe the linkage formed between papers which have references in common. Two papers are said to be bibliographically coupled with coupling strength of one (two, three etc.) when they have one (two, three etc.) references in common. Kessler has found that groups of papers coupled in this way are quite similar to the groups of papers coupled in this way are quite similar to the groups made by a conventional subject index.

Indexing by means of bibliographic coupling is likely to become more common in future than it has been in the past. Relying on the authors of papers for the intellectual efforts which is required for a good index, it demands the use of new search strategies on the part of library users. The technique is particularly suitable for machine applications in that no intellectual effort is required to construct the index. The printed volumes of Science Citation Index are compiled on these principles and make it possible to experiment now with search strategies which will be appropriate with some future on-line information retrieval systems.

The automation of library operations can very readily be regarded as a way of escape from the pressing problems of everyday life in a library; ordering and cataloguing books, answering awkward reference questions, tracing lost books. Automation will certainly simplify many of these tasks, but it must be built upon a sure foundation of good library practice which, as far as possible, prevents problems from arising.

Librarians best know what the problems are in cataloguing, acquisitions work, reader service, and circulation control, and it is important that their knowledge be brought fully to hear on the job of designing automated systems in the first place. Librarians must understand fully what automation can do in all areas from book ordering to natural-language text processing. Then this knowledge and experience will enable the goals of automation to be defined rightly and attained more readily.

A computer is an electronic machine for manipulating symbols. For out purposes, we can readily make this definition more meaningful if we say that a computer can be regarded in many ways as being like a clerk. And if it is a clerk, then we may be able to use it to do clerical work in libraries more quickly, accurately, and more efficiently than it is done at present. It might even be possible to teach it some professional or intellectual jobs. Being a machine it should not tire under heavy work loads, and in fact, should become more efficient as the work load increases so that as library grows this can be accommodated without proportionate increases in staff. From what we know about computer applications in other fields, we should expect automation to be feasible only in larger libraries or in cases where a number of smaller libraries combine together to form one large unit for processing purposes.

However, we know also that computer time is growing steadily chapter as newer, more advanced machines are devalued, whereas the time of human clerks is becoming steadily more expensive. So we predict that automation will gradually became feasible in smaller libraries. This trend will be accentuated somewhat by the emergence of very small, cheap computers, but it must be remembered that many library applications require a large machine. The growth of on-line techniques, which are particularly suited to some areas of library work, will make it realistic for libraries to expect to have on-line access to a large and powerful machine, rather than consider purchasing a small machine for their own use.

A well-established set of procedures exists, leading from the point we have now reached, that of having stated n general terms the advantages which might be expected to accrue from using a computer, to that of installing, testing, and running an automated system in a library. To follow all these procedures faithfully might seem pedantic, perhaps unnecessary in many circumstances, but it is necessary if the job is to be done honestly, avoiding emotion, cultivating reason, and working in quantitative terms. Because procedures will be enumerated and discussed in one particular way, it does not mean that this is the only way, or that different combinations and distributions of these procedures might not result in very different patterns of overall working; only that the basic ideas must follow one another in this way.

The first step towards introducing an automated system is to carry out a feasibility study. This may be done at any level of detail, but the purpose is to ascertain whether or not a computer can be used to do a particular job more efficiently than heretofore. It can be an expansive stage, and may be broken down into two parts, the first one to ascertain if the feasibility study itself is worth undertaking. It is common knowledge that many such studies have shown that a computer was not ledge that many such studies have shown that a computer was not really needed at all, and that the information gained from the feasibility study was sufficient to improve the existing system so as to make it more efficient than the automated one first contemplated. It is during the feasibility study that systems analysis takes place.

The system, be it the whole library itself or only one department, is analysed; its method of working is systematically taken apart piece by piece, so that in the end all the pieces are known and understood, as are the ways in which they fit together to make the whole. If only one department of the library is being investigated, the relationships between it and the other departments, and the library itself, must also be known and understood. Measurement will also take place so that the final result of the study can be in quantitative terms. Numbers will be counted; the numbers of items bought, catalogued, lost, circulated, damaged, and the way these number may be expected to change in the future will be discovered.

Also counted will be the number of people needed to do the various jobs in the system, and the time to do each jobs in the system, and the time to do each job will be measured. It is very important that the end results of all this library activity be calculated or estimated. This can be difficult because it is not easy to apply a measure to a student reading a book, a chemist pleased with a morsel of information, or a housewife experimenting with new dishes described in a cookery book borrowed from her local public library. It must be attempted, however, because any automated system, as we have seen, is likely to alter the pattern of service offered by the library, and this is as important in the final evaluation as the numbers of books ordered each week, or the time it takes to discharge an overdue book.

From all these measurements the costs of the present system will be calculated. They will almost certainly be expressed as unit costs—the cost of cataloguing the "average" book, or answering the "typical" reference question. Together with details of the services performed by the present system and in particular any difficulties and problems associated with it, any aspects which are like to become unsatisfactory in the light of future developments will be indicated, and all will be used by management to make an educated decision about the introduction of automation.

Because the introduction of computerised operations is not the only, and not necessarily the best solution to the problems that will have been brought to light, the next stage of the study is just to find the best solution, automated or not. It is highly desirable, therefore, that the study should be carried out by persons who have no direct interests in seeing that a computer system is installed, perhaps even using a particular make of computer. They should be perfectly free to advise on the optimal course of action. Like wise, the investigators should have considerable experience of similar studies, and be well informed on all data processing and work study techniques. Needless to say there is a great shortage of such people. Feasibility studies, therefore, are not cheap and so are often skimped or omitted entirely.

If the introduction of automation is indicated, and this is the only case that really concerns us here, then the second stage of the study will involve the selection of a computer and the various necessary ancillary items of equipment and stationery needed by the system. There are considerable differences between the different makes and metals of computer on the market, so that the selection of the machine most suited to the purpose on hand will have a marked effect on the final operating efficiency of the system.

The end result of this second stage of the study is a broad system specification. It will indicate the hardware or machinery required and, in general terms, how everything will work. The through put or work load handled will be given, as will the estimated costs. Staffing requirements will almost certainly be different from those of the old system, for keypunchers will replace clerks (a clerk can be trained to keypunch, of course) and it will be necessary to have a senior staff member, with data processing experience, in charge of operations and one or two programmes may be required on a permanent basis. Staff requirements will be very different where a library installs a computer for its own use, from the case where it merely uses a short time each week or someone else's machine.

Because computers work in ways radically different from human beings, the flow of work in an automated system can differ quite considerably from that in the corresponding manual system. Old departmental barriers can therefore be broken down of the introduction of automation, as similar processes in different departments are combined and as duplication of work is eliminated. The overall organisational structures of the library can be materially altered, and this will be appreciated and stated in the feasibility study.Once an automated system is considered in any detail it is apparent that it offers a range of possible products and services considerable winter than does a conventional system. The compilation of all kinds of statistics is but one example of this.

A library can expect, therefore, to expand the scope of its operations by introducing mechanised routines, and it is desirable either that the team doing the study include a good knowledge of library service among their skills, or that one member of the library staff work closely with the team, in order that the full possibilities of winter operations are realised. From the final report of the feasibility study it should be a comparatively simple matter for the library management to decide on a course of action. Only in the largest libraries would a feasibility study be likely to indicate clear overall cost reductions through introducing automated methods.

In other libraries the situation would almost certainly be that at the present time a computer-aided system would cost more but would bring additional advantages. It is up to the management to decide whether these advantages justify the increased costs, bearing in mind that, in time, costs will tend to change in favour of automation. If we assume that a decision is made in favour of automation, where does the library go from here? The answer is that the next stages in this idealised introduction of mechanisation will depend very considerably on the extent of the automation which is planned, on the organisational framework of which the library is a part, and no what we can term the library's computing environment.

The procedures for implementing an automated circulation system, for example, will be different from those required by a system embracing all a library's technical processing routines. In a university environment the library would expect to make use of a central university computing facility rather than purchase its own machine or use the services of a commercial bureau. This will affect not only the type of machine used, which is likely to be better suited to mathematical than to library work, and the languages programs are written in, but also, because programming may well be done in the computing laboratory, it will affect the staffing requirements in the library.

The overall computing policy of the organisation of which any library is a part will also affect such factors as, for example, the frequency with which computer runs can be made, or whether on-line working is, or is planned to be, available, and so be an important factor in the design of the library system. In large organisations it not infrequently happens that a library in another division or department has already automated some of its functions. It may be possible to take these programs over just as they stand or with only small modifications. In the majority of instances, however, the library's computer system will have to be created from scratch. This stage—system design—can be regarded as consisting of two main parts, although in practice they can so merge together as to be indistinguishable. The first part involves floe-charting the ways in which the system will work, and then in more detail, the several operations and decisions performed by the computer in working the system.

The second part, which is the job of programming, carries on in more detail and converts the flow charts into a sequence of computer instructions—a program. It is quite possible for one person, or one group of persons, to carry out both of these two steps, in which case the distinction between them is obscured. Alternatively, the second step, which has more of the character of straight forward routine than the first, can be performed by a programmer.

In either case it is absolutely vital that at the stage of system design proper (step one) all the implications for the library of the various detailed decisions and operations which will be included in the final program are fully understood. This requirement can be achieved, as during the feasibility study, by having as senior library staff member work with the system design group, but, ideally, the system designer or designers should themselves be experienced librarians. Similarly, it is best for the programming to be done by someone with a good knowledge of librarianship.

Few people, even nowadays when their great worth is appreciated, are at the same time experienced librarians and knowledgeable and up to date in computing matters. If it is vital that an automated library system be designed with all its implications for librarianship in mind, it is, at the very least, highly desirable that the programs are written so as to make the system as efficient as possible from a computing point of view. So library systems are very often designed as a co-operative venture between library and computing department staff in order to assemble together the necessary skills and expertise.

The work is often done as a sideline, or even in staffs' own time. That so much sound work has been more rapid and more quantitative analyses made had these methods of working been less common. Where it is difficult to find the time to design and implement automated system, the detailed analyses and comparisons necessary for its full justification and evaluation naturally tend to be skimped or omitted. Overall impressions are substituted and, as a result, the library profession wherein the arguments of emotion can flourish.

The languages in which computer programs for library routines are written are important subjects for discussion at the present time. For mathematical, commercial, and other special kinds of work, a number of high-level programming languages have been defined. These make it easy for the programmer because the programming language can be quite similar in form and content to the language ordinarily used in the particular discipline. The price paid for this programming facility is a less efficient use of machine time. This is unimportant in one-off jobs, but it becomes a major factor in library systems where a program is run frequently.

A further advantage of high-level languages is that they can often be run on many different makes and types of computer; on all machines, in fact, for which a special translating or compiler program has been written. This takes each instruction in the high-level language and converts it into the necessary number of detailed machine instructions for the computer concerned. Programs written in high-level languages can thus be exchanged widely, to the overall conservation of programming manpower.

Because of the machine inefficiency of high-level languages, library programs tend to be written in what are termed low-level languages. These make better use of computer time but, being less similar to natural language and requiring the use of quite detailed instructions, make greater demands on the programmer's time and ability. They are generally applicable to only make or series of computers, and so do not permit free exchange and utilisation of programs.

In a situation where computing power is at a premium, library programs will tend to be written in low-level languages at the expense of programming manpower. Where programming manpower is at a premium, the tendency will be to use high-level languages at the expense of computing power. Present trends are beginning to indicate that, in future, library programs will be written in high-level languages and exchanged freely in order to save the very considerable effort which would be entailed were each library top design and program its own systems.

When a program, has finally been written, it must then be tested, or "de-bugged", to remove as many as possible of the errors which the fallible human system-designers and programmers will have written into it. This is generally done by running the programs with test date to simulate as many as possible of the circumstances that are likely to arise in real-life working. It is at this stage that the value of an experienced programmer is most apparent. He will have written fewer errors into the program to start with and will e able to discover more readily why things go wrong when they do, and be able to put them right. Programming skill in this respect is particularly valuable when working with low-level languages.

After this stage the program is ready to being work with real library data, nut it is usual to place one more safeguard against error in the machine system by working the new system in parallel with the old for a few months. After this period, if everything is satisfactory, the support provided by the old system in event of failure of the new is removed and the automated system, having been thoroughly tried, takes over full responsibility. Machines, of course, like human beings, are not infallible and they do from time to time make mistakes or break down.

A "safety net" must therefore be provided so that in the event of machine failure no data is lost and the system can carry on, or safely suspend operations until the failure can be rectified. Such a safely suspend operations until the failure can be rectified. Such a safety net can be provide with a coarse or with a fine mesh, the latter being more expensive. A coarse-meshed net, commonly termed a "fail soft" sub-system, permits only limited degree of confusion and disruption of operations to result from machine failure. A net with fine mesh, a "fail safe" sub-system, constitutes an analogous set of producers with allow operations to continue normally with, hopefully, no confusion. The sort of net employed in each situation will be determined by the importance of the particular system and the value to be placed on keeping the relevant records intact.

Once an automated library system is working satisfactorily, it becomes important that it should continue to do so. Procedures in the event of machine failure are only one aspect of this general problem. Equally important is the need to establish a good working relationship between the library and the computing department so that the desired standard of library service is maintained. A library system usually demands that a computer run be made at a stated time each day, week, or month. For example, where a university library has introduced a computerised circulation system, it is necessary that an updated printout of books on loan should be available in the circulation department each day. If anything cases this printout to be delayed, then the work of the circulation department will be hindered.

In fact, in the case of circulation control it is true to say that the whole justification of an automated system hinges on the daily printout being available at the right time each day. Library work must often therefore receive high priority in the computing department, and this can be difficult to achieve. In some organisations, however, computing policy, and thus the order of priority of the different users, is determined by a special computing committee and not by the head of the computing department. If the library can put a sufficiently strong case to such a committee, then the necessary degree of priority will be guaranteed and, consequently, the desired quality of library service ensured.

In some instances the first move towards library automation comes from within the computing department, and then this difficulty should not arise if the importance of priorities is emphasised by the librarian from the early stages of the project. The correct note will be sounded, if, instead of speaking simply about "changes" in computer science, we use the word "improvement" and say that improvement are continually and very rapidly being made to computers and the ways in which they can be used. This it is certain that within a few years of introducing a particular automated system, improvements in computers will make it possible to re-design the system for a more modern machine so as to extend the scope of what is being done and to do it more cheaply.

Of course it may well happen that the particular computer used by the library will not have been replaced so quickly and improvements which would be made possible by a newer machine cannot be made. It is unusual for a given computer to remain in service for more than 10 years, and after 5 years newer machines will being to make a change very desirable.

The implications for libraries of this state of flux are very considerable, especially since, until recently, libraries have exhibited only a very slow rate of change. Librarians, despite or perhaps because of the conservations they display in relation to changes in libraries, show a remarkable interest in the very latest computing techniques that are either still in the early experimental stages or no more than ideas in designers' brains. These future developments invariably hold out the promise of solving all the world's problems of bibliographic organisation, and, perhaps not unnaturally, librarians grasp eagerly at them. But when the full implications of such developments are pointed out in terms of the sacrifices and changes each library would have to make in order to adopt them, then the former panacea can become less attractive.

Once a program has been implemented in a library it will not be long before the need will become apparent to make small additions and changes to it. These must be done by a programmer and will sometimes involve considerable alterations in the whole structure of the program. Where automation in a library has progressed beyond an elementary stage, it will become very desirable for the library to have a programmer on its permanent staff and perhaps a more senior person, a systems analyst, as well.

An embryonic data processing department will have been created within the library because of this requirement for constant minor change and improvement. It will have responsibilities in all the conventional library departments where automated routines have been introduced and must have the authority to see that any changes are made in such a way that they satisfy the requirements of efficiency laid down for the system itself.

Once such as department of data processing has been established, a library will be in a good position to keep abreast of advance in computer science and will be able to modify the library's own systems from time to time to take fullest advantage of changes in the facilities offered in the computing unit where the library's work is done. Being actually on the library's own permanent staff, members of the library data processing department will be able to take full account of the librarianship aspects of their work. This is the ideal situation, and where a library is introducing automation on anything more than a very limited scale, the establishment of such a department should be given high priority.

In order to complete this picture of libraries and automation it is necessary to try to look a short distance into the future, to see what development are likely to come about in computing and how libraries will be able to use them. Because of the rapid rate of change in the field of automations, such future knowledge becomes essential if one is fully to understand the present situation and to know how to plan for library systems which will become operational in a few years' time.

There is one fairly recent development in this which should be of the greatest importance in the future. And because here is something which is with us now, holding potentially great possibilities for the future, it is possible to speak with some certainty. There now exist in Britain a number of working parties, each active in some areas of the field of library automation. Each owes its existence to the fact that a number of library automation. Each owes its existence to the fact that a number of libraries are co-operating and pooling their resources to speed progress. Some of the working parties are designing systems for a particular make of computer, while others have no such limitation. Some are concentrating on the problems of one particular type of library, while others consider all types of library.

It is beginning to seem likely that when these groups finally produce viable systems each will be adopted with little or no change by several libraries. One group in particular is designing a book-ordering system in such a way that it includes most of the features required by libraries and can readily be altered to cater for special individual preferences. This co-operation will be of the greatest importance in future. Now, and always, it makes possible savings in programming effort and speeds the introduction of automation where this has been shown to be justified. When cataloguing data in machineable from become more generally available, for example, its use by individual libraries will be facilitated where numbers of libraries are using identical or similar cataloguing systems.

The situation will be similar where a new technique or piece of equipment becomes available to a group of libraries which are using one type of computer. A new improved program need be written once only to be capable of adoption by all members of the group. Once a group of libraries have automated all or some of their activities and have proved this to be an advantage, other libraries will find it to their advantage either to adopt the common systems as they stand or with the minimum possible alterations. Among university libraries at least, programs will be made available freely.

It is much easier to talk of library co-operation in this way then to achieve it in practice. Although all libraries are the same to the extent that they buy, catalogue, store, and lend books and other material, they do differ considerably in that they buy different kinds of books, catalogue them in different ways, and serve many different types of readers. Accordingly, libraries have developed in many different ways, and rightly so. But it is nonetheless true that when comparing the various methods of performing a given job, the differences tend to be in matters of detail, and that underneath is a large area of common practice. This makes it reasonable to expect that co-operating libraries will be able to accept large segments of automated systems as they stand and will need to later only details. It is true that some apparently trivial differences in points of detail can make the adoption of a whole system impossible, but such cases are the exception rather than the rule.

There is a very broad spectrum of different types of library, each with different needs, but it should be possible for the whole spectrum of needs to be satisfied by there or four major types of automated system. Three such divisions of the spectrum of libraries might be as follows:

(a) Small libraries, having simple cataloguing needs and buying straightforward material from limited number of suppliers.

(b) Medium and large libraries, buying a wide range of materials from all parts of the world, with cataloguing needs more complex than type (a) but not so demanding as (c). Full control of circulation s needed and individual treatment, particularly as regards cataloguing, is required for special collections. Most university and public libraries would fall into this class.

(c) National and other very large libraries with complex acquisition procedures for all types of library materials. Fully detailed cataloguing is necessary for very large catalogues which may have to be related to the production of a national bibliography.

This tendency towards standardisation is, of course, to be observed in most aspects of life today. Its advantage is to be found in the reduction of costs it makes possible. Despite our understandable attachment to our differences, we accept standardi-sation because of the clear advantages it offers.

A good illustration for our purpose can be seen in the motor industry. People's transport need tend to fall into a small number of well-defined classes, and so do the types of motor car that are produced. A fair amount of individual differences in needs and preferences are catered for by different manufactures and by the addition of optional extras and preferences can be catered for by adding "optional extras" to the above three design types, and further variety will be ensured by the fact that differing systems will be produced by the various working parties mentioned above.

Included in this concept of standardisation of library systems is greater uniformity in cataloguing practice and subject identification. A small number of universal schemes for classification and subject indexing will be adopted by more and more libraries. These universal schemes are with us already and, despite their acknowledged short-comings, libraries have already begun to change over to using them. The advantages to be gained from a central source are considered to make worthwhile the very great task of conversion involved, for example, in reclassifying a library's whole collection.

The greatest pressures towards standardisation will come in this area of cataloguing as a result of advances in the practice of librarianship that are associated with the introduction of automation. Three events are combining together at the present time which are potentially capable of effecting very great changes in library practice throughout the world. Firstly, the Library of Congress Shared Cataloging Program is extending the scope of the Library of Congress printed catalogue card service to include all material from western European countries. Secondly, the introduction of standard book numbering in Britain, besides facilitating the acquisition of books by libraries, provides a unique tag for each book which can readily be handled by computer, and enables cataloguing copy for new books to be recorded by the British National Bibliography from 3 to 6 weeks before publication.

If standard book numbering is adopted by other countries, and there is good reason to hope that it will, be then the cataloguing copy provided by the Library of Congress shared Cataloging Program will also be available a similar period before publication. The third event, or rather programme of events, is the Library of Congress MARC (Machine-Readable Cataloging) Project. When this has been fully implemented it will make the cataloguing copy, collected in advance of publication by the Shared Cataloging Program, available to libraries in machineable form. The format of these machineable catalogue records is designed to facilitate the exchange of cataloguing copy between libraries and to enable individual libraries to modify and adapt the records to their own particular needs.

Such modification can be of two kinds, programmable and non-programmable. Programmable alterations can be performed automatically by the computer when the machineable catalogue record is added to the library's own internal records. They included standard abbreviating operations on headings, descriptions, and classification number, and the reformatting of entries. The basic idea behind such programmable alternations is that they are standard. Non-programmable alterations involve human intervention and are needed whenever part of a catalogue record must be changed to some form that cannot be derived by algorithm from the original record. All such alterations cost money and the non-programmable ones will be the more expensive of the two kinds.

Centrally produced cataloguing copy, of course, has been with us for some time, and it has been bought by many libraries, especially in the United States. Two drawbacks have prevented its more wide-spread use; the comparative slowness of the service and the low percentage of libraries' total acquisitions covered. Both these disadvantages should expect to have approximately 70-80 per cent of their new acquisition covered by the Library of Congress service. Librarians have long realised the advantages of central cataloging service and have been aware of the great waste of effort when books are catalogued in each library where copies are bought. Now it is likely that this long-standing dream will soon materialise.

Another librarians' drams may also be realised over the next few decades as a result of this co-operation in cataloguing : the national catalogue given the holdings of all libraries in the country and also containing entries for all new books as they are published, able to be searched by computer from remote on-line terminals, appears to librarians to solve very many of the problems that they have today.

There are very great obstacles to be overcome before such a catalogue can be made a reality, but work being done now leads one to expect that it will eventually come to exist. Machineable copy for all new books it almost with us and the investigations now in progress 9 on how to convert the entries in the British Museum catalogue to machineable from should enable a very large corpus of existing catalogue material to be included in the national catalogue. The very real difficulties in programming a machine catalogue search by author and title, cased by inconsistencies in cataloguing, can be overcome by standard book numbers for new books and for old ones. Direct access computer memories are rapidly increasing in size and should cause no serious problems. The remote on-line techniques are already with us.

The advantages of such a catalogue are considerable. Primarily, it will give the locations of material listed in it and remove almost all the inefficiency and frustration from existing systems of inter-library lending. As such, it will be a major adjunct to national scholarship. Libraries will be able to extract cataloguing data for new acquisitions directly from it by means of on-line consoles. This act will automatically indicate a location in that library for the items concerned. Local catalogues, whether printed on paper or kept only in machine memory, will list the most frequently used material only. Thus, great savings will be realised in catalogue printing costs or in machine memory storage devoted permanently to library catalogues. Uniformity in cataloguing practice will be further encourage. It is questionable, however, whether such a catalogue will be searchable by subject.

From this point on, libraries' dreams of the future begin to bring the handling of textual material within the scope of systems of automation. Facsimile transmission of photographs is an everyday occurrence in the newspaper industry where speed is all important and where the volume of information is comparatively small. Transmission of printed pages by these means is practicable and economical over short distances where there is a high traffic density, but up to the present there has been little demand from libraries for higher transmission speeds than those provided by sending photocopies through the post.

Prophecies of libraries where the codex and the printed page have been replaced by digital storage devices and where a computing facility has taken over the task of literature searching is very exciting and attractive, both to librarians and to those scholars who really are overwhelmed by their century's information explosion. Although many people would regret the passing of the codex and although the combination of the printed page, the codex, and the library is embarrassingly efficient as a machine for storing data and providing rapid random access to it, there is no doubt that many research workers would welcome real assistance with their information gathering problems.

Good assistance would be provided by a skilled librarian, but in a world where skilled people are in short supply it is likely that some machine system will eventually take over this task The techniques that such a machine would use may be quite different from those now used in libraries, but the end result would be the same. Our task as librarians is both to design these techniques and to see that we do not hinder their adoption, where appropriate, by arguing from emotion. One such technique, now with us, may be that of bibliographic coupling, and certainly the methods now being developed for machine processing of natural language texts will have an important role to play in the operation of this future automated library.

Effects of Library Automation

Automation has affected most library processes to some extent, and in many cases has influenced the development of library policy. In the area of cataloguing, however, the effects of library automation have been so far-reaching and profound that a separate discussion of their impact seems warranted. The effects may be seen in four general areas:

— Extent of cataloguing information presented to the users

— Choice of information given

— Arrangement of information

— Typographical presentation of such information

Typographical presentation of cataloguing information

The last of these areas is one of the most interesting from a mechanical point of view, but certainly the least significant so far as cataloguing policy is concerned. Librarians, however, have bemoaned the "uppercase mentality" of computers and related machines since their use in library work began. The resulting catalogues were certainly less attractive and less readable than the printed or typed cards with which most people were familiar. Within a few months of Clapp's talk, however, the situation began to change. It did cost a substantial amount in computer processing time, however, which led to its abandonment by Atlantic University.

Photocomposition machines, which did not share this handicap, later became available, and in some cases offered the additional advantage of a wider variety of type faces and images than older printing techniques (the Videocomp and similar machines, for example, can compose and print Oriental text, symbols of all types, and if fact any image that has first been converted to electronic from using a special process). Microform catalogues and computer terminal displays have encountered the same problem, but both have now advanced at least to the stage that acceptable upper and lower case characters in the Roman alphabet are available. Non-Roman alphabets and nonalphabetic characters are still a problem, but the capacity for providing these exists if the demand becomes sufficient.

Changes in the extent of cataloguing information presented to the user under each entry have been mainly deletions rather than additions, and such changes are most frequently seen in book catalogue projects. Virtually every descriptive element has been eliminated or abbreviated in one project or another. The Boeing SLIP catalogue restricts author's forenames to one and represents the others by initials; other catalogues give only initials, with no forenames. The University of Rochester catalogs limit the main entry to twenty-one spaces, with necessitates a substantial amount of abbreviation in corporate authors.

Collation is perhaps the most frequently abridged set of elements. Standford and Florida Atlantic give pagination only, and many catalogs give no collation at all. Notes and tracings are also frequently omitted, with the exception of series and contents notes in a few cases. The ONULP (Ontario New University Libraries Project) catalogs, however give full tracings under main entries, as does the National Library of Medicine Current Catalog, the latter because other medical libraries use the catalogue as a source of cataloguing information.

In many catalogues the amount and type of abridgement varies according to the type of entry involved. Normally the main entry is reasonably full, and most abbreviation or abridgement takes place in added entries. In the Los Angeles County catalogue, however, the subject entries are more complete than the author entries, because the subject entries include an annotation.

The practice of abridging entries has been justified on several grounds. Undoubtedly it began, especially in catalogs produced with unit record equipment such as the University of Rochester listings, because of the eight-column limitation of punched cards and the printing limitations of tabulators. As Weinstein and George have noted, "the one-to-one, punched card/ printed line relationship, and line-length/field-length compromises, all encourage brevity of entry.

At least in part, however, the elimination of data elements also reflect a desire to cut costs. Simonton has viewed it "as a reflection of the traditional policy of many of these libraries of shortening the bibliographical record for more economical processing."Another justification is based on the shorter life expectancy of the book catalogue as compared to the traditional card catalogue:

> As [card] catalogues serve as the basis for the permanent record and their cost can be prorated over several decades the need for a careful description of the many facets of a book is quite properly justified. In the case of a book is quite properly justified. In the case of catalogue supplements, however, where the record will serve quite likely for only a few months, any attempt at detailed description of the book cannot be justified.

As might perhaps be expected, the old argument over whether the catalogue serves as a bibliographical tool or a "finding list" has also been revised. Robinson, for example, writes about the planing for the Baltimore Country book catalogue project:

> We also agreed on the concept that this catalogue was to be a finding list, not a bibliographical tool. We had always done fairly simplified cataloguing anyway, and frankly, we were convinced that, as bibliographical tools, catalogue are vastly overrated and underused anyway. Taking an educated guess, we estimated that only 50 percent of the people who come into out libraries use the catalogue at all, and 98 percent of those use it as a finding list.

Not all changes in the extent of cataloguing information presented have been abridgements or deletions, however; some writers and projects, in fact, have seen automation as a means of increasing the amount of information made available.

In the MARC project, which was developed soon after, the items mentioned in the first paragraph above were indeed made explicit. MARC has also provided additional "access points," as indicates, although not chapter headings or additional name and subject entries as Avram and Markuson suggested. Many catalogs of special libraries have provided additional subject headings, however, taking advantage of the ease with which machines can reproduce entries under various headings. Most book catalogs also reproduce title entries for all items in the catalog, not just those with "distinctive" titles, and as mentioned in the discussion of the Boeing SLIP catalogue some also "permute" the titles so that they appear several times for each work, under each significant word in each title.

The choice of main entry has long been a critical issue in cataloguing policy and in the development of cataloguing rules, but with the advent of machine-produced book catalogues the need for the concept of "main entry" may be questioned. When catalogue cards were printed or typed by hand, it made economic sense to produce only one card, with the data arranged in a standard format under a "main entry," and then to reproduce other cards as needed, typing only the added headings required after the "unit cards" had been duplicated. Computers, however, can rearrange data in any fashion deemed appropriate, so this rationale is no longer applicable. Despite this fact, many book catalogs continue to follow the unit card and main entry concepts. The ONULP catalogues. For example, print the author under each title entry, and then repeat the title a practice that would appear to have no justification except for mechanically reproduced catalogue cards.

A few book, catalogue projects have abandoned this anachronistic practice and reproduce entries under various headings without any attempt to designate one as the "main" entry, even though the entries under each heading are still "unit" (i.e., identical) entries. Abandonment of the main entry concept, it should be pointed out, also means abandoning and attempt to designate the person or body primarily responsible for the intellectual content of a work, but clearly it makes the choice of entry less critical, particularly in difficult case such as conference proceedings, where the choice between conference title, sponsoring body or editor may be little more than arbitrary.

Arrangement of catalogue information

One of the greatest effects of automation on cataloguing practice to data has been in the arrangement of entries in relation to each other. Virtually all book catalogs are divided catalogs, abandoning the so-called dictionary arrangement that interfiles all types of entries in a single sequence.

The type of division is not consistent from one catalogue to another. Some have separate author, title, and subject sections; some combine author and title entries in one catalogue with a separate subject section; and at least one combines titles and subjects, with the author entries separate. Regardless of the method of division, however, two principal reasons are usually given for the arrangement: users make different uses of subject entries as opposed to author and title entries, according to a number of user studies, and the filing or sequencing of entries can be simplified if all types of entries are not interfiled. The latter reason is normally the more urgent when automation is involved. Catalogues can be divided without being automated, and if ease of use were the only justification for division there would be no reason to attribute the popularity of the practice to automation. Filing, however, is another matter, and simplification of filing for automation purposes is undoubtedly the main reason for the predominance of the practice.

Computers can sequence and arrange data with extreme speed and accuracy if given explicit rules for doing so, but traditional library filing arrangements are a mixture of alphabetical, chronological and other sequences presumed to be "logical," with numerous general and specific exceptions. In addition to the complexity of the rules, many of them are based on the ability of human filers to recognize information that is only implicit, not explicitly stated, and computers can work only with explicit data. At least five categories of difficulty may be identified:

— Information in the headings must be ignored or removed in order fore computers to sort it acceptably.

Honorific titles such as "Sir" in proper names are also commonly disregarded, so that "Drake, Sir Francis" files as "Drake, Francis."

— Information that is implicit must be made explicit. A common example is "U.S.—History—Civil War," which as a subdivision of "U.S.—History" files not alphabetically under "C" but chronologically, as if it were "U.S.—History—1861-1865." This category of problems is perhaps the most difficult for computers to deal with because there are no clues for most such instances, sand the number of "as if" situations in most current filing practices is large. It is also the least defensible. "Dr." is filed as if were spelled "Doctor" or "Doktor," depending on the language of the text, nut if the reader does not know the rule, or has only a title such as "Dr. Jacob Dubs" to go on, he may miss the entry entirely. "Mrs.," (and presumably "Mac," although McKenzies insist they are not Mackenzies. Numbers are a special example of the "as if" problem; they file as if spelled out (again depending on the language of the text) at the beginning of titles, but file as numbers, in numerical sequence, in headings for numbered conferences, series and so forth.

— Information in headings must be rearranged in order for machines top sort it into the desired sequence. "Charles II, King of England" must be read as 'charles, King of England, II"—or actually, since there is also a numeral in this case and the second category of difficulty comes into play, as "Charles, King of England, 2."

— Exceptions must be made to alphabetic order. Some rules require that entries beginning with the same word or phrase be sub-arranged not alphabetically by the next filing element but by bibliographic function (authors, then subjects, then titles, for example), or according to the type of entity the heading describes (persons), then places, then things).

— Punctuation is even a problem, because of complications introduced both by the machines and by filing rules. Computers regard punctuation marks as character just as letters and numerals, and sort them into a certain order (differing to some extent by brand of machine) unless programmed to do otherwise. Extra spaces inadvertently inserted by humans will affect machine filing order as well. Existing rules differ on the treatment of certain marks, for example, whether "Brewer's Dictionary of Phrase and Fable" should be filed as "Brewers" or "Brewers," and "Mott-Smith" as "Mott Smith" or "MottSmith."

The ideal solution to these various problems is unclear, but attempts to solve them in automated catalogs have produced a significant amount of change as compared with manually filed catalogs. Most projects have used a combination of four basic approaches:

— Continue to follow present filing rules, using people to assign additional codes to headings as necessary to allow computers to arrange the headings as desired.

— Abandon present filing rules, and use the arrangements produced by the computer's sorting sequences with a minimum amount of change.

— Change the headings themselves as necessary so that computers can arrange them in accordance with very simple filing rules.

— Attempt to program computers so that they will arrange headings automatically, either by present filing rules or by modifications of them.

Florida Atlantic University used the first approach in its book catalog, the first edition of which was published in 1964. "Nonfile" symbols were used to indicate to the computer that portions of an entry were to be ignored in filing. Determination of when the symbols were to be added and actually adding them were of course operations performed by human beings.

Use of the symbols was limited to initial articles, qualifying phrases such as "jt. author," and titles of honor; for example, "Churchill, Sir Winston," was input as "Churchill, Sir Winston," to enable it to sort as "Churchill, Winston." Umlauted letters were converted to letters without the umlaut, but followed by an "e." Unfortunately, this also meant that "cooperate" became "cooeperate," and "Bronte" became "Brontee." Abbreviations were filed as written, not as if spelled out, and "Mc" was filed as "Mc."

A considerable amount of manual coding was performed in the University's main card catalog. Entries were sorted not on the text of the entries themselves but on separate "sort keys" composed of a total of 100 characters, 80 from the filing element (the author, added entries or subject headings) and 20 additional characters from the title, which was used throughout the catalogue for subarrangement of entries under each heading. An obvious disadvantage of extended use of sorting and filing codes was noted at Stanford: the technique requires manual maintenance of authority files, very careful inputting, and very careful editing and proofreading.

Few projects have attempted the second alternative (following the computer's own sequencing arrangements), because of the radical differences from traditional filing that would result, but a number of catalogs have used the third approach—changing the headings so the computer would arrange them in accordance with simplified but more or less traditional filing rules. Usually such projects have taken a strictly pragmatic approach, solving the simplest problems and ignoring the others. The Baltimore County book catalog, first published in 1965, is an example. Initial articles, titles of honor and such designations as "ed." or "comp." were omitted from headings during input, which of course meant that they were ignored not only in filing but also in printing. Abbreviations, numerals, and symbols such as the ampersand were written out. Some titles were rearranged, e.g., titles of annual reports were input as "annual report," followed by the date. In all other cases, the computer sequencing was accepted as it came.

The foremost advocates of changing headings to produce the desired sequence have been Theodore Hines and Jessica Harris. In 1963, Hines urged that "filing should be a purely mechanical operation which can be reduced to a straight forward arrangement of sorts and nuls. The filer or program should not be expected to expand or interpret entries for filing purposes. These functions should be shifted to the formation of entries."

Publication of the Hines-Harris code created a storm of protest. One reviewer warned, "This little book seems likely to do more harm than good. The end it proposes may be laudable, but the means are not. The book rests on the proposition that what is awkward for the machines must be eliminated and the simplicities of the machine system imposed on non-machine processes." Another criticized the proposal for failing to make efficient use of the computer: "Entries are arranged only after the format has been decided by a human.... If we must spend professional time in this manner, what, then, is the advantage of using the computer?"

The fourth approach, which is the extreme—attempting to program the computer to follow traditional rules as closely as possible— was taken by William R. Nugent of Inforonics, Inc., in a study commissioned by the Council on Library. Resources. Unlike most proposals which advocated sorting abbreviations and similar exceptional cases "as is" rather than "as if," Nugent proposed a system of "substitution notes" which would continue traditional "as if" filing. Extensive rules to permit arrangements by category (under Bible entries, anonymous classics, and corporate authors, for example) were also suggested. In not every case, however, was Nugent happy with his assignment to propose ways of "programming the LC Filing Rules."

> If one set out deliberately to destroy the simple precision of numerical hierarchy, little worse could be done than to insist that numbers be ordered by the spelled out equivalents of their spoken sound. In this writer's view, the present filing rule should be quickly replaced by one in which all cardinal or ordinal numbers, whether in alphabetic or numeric from, would be ordered by their elementary numeric from.... As this is unlikely to occur with any haste, it will probably be necessary to write programs to covert the numeric portions of entries to alphabetic from.

He then pointed out that it would be necessary to devise further rules to ensure consistency, because "1810" could be written out as "eighteen ten," "one thousand eight hundred ten" or "one eight one and the existing LC Rules convert such numbers "to differing forms by application of some unspecified and doubtless unprogrammable subtlety." This kind of problem of necessity led Nugent to propose various types of manual coding as well as elaborate programming.

Perhaps the most thoroughgoing and thoughtful consideration of the entire range of automated filing problems is in a series of papers by John C. Rather, Chief of the Technical Process Research Office in the Library of Congress. In February 1970 Rather distributed a working paper entitled Filing Arrangement in the Library of Congress Catalogs within LC and to a limited extent outside it, setting forth a series of principles and basic approaches. Consensus within the Library led to public distribution of a full set of rules with the same title the following year.

In 1972, the principles and basic approaches were elaborated in print, accompanied by an abridged version of the rules. They have since been used as the basis for a computer program to arrange MARC records and produce a variety of book catalogs for LC. Since they are also being considered for using in automatic preparation of the National Union Catalog, their significance is obvious. Rather is eloquent on the dangers of changing headings merely because of the filing difficulties they present in automated systems.

> The complexities of catalogue headings and their arrangement are not attributable primarily to the rules that govern them; they arise from the varieties of materials being cataloged and the size of the catalogue being maintained. The need to differentiate names in the catalogue while still conveying intelligible information about the name result in forms of heading that indubitably present difficulties of arrangement. The fundamental conflict between the function of headings as units of intelligence and their function as filing guides cannot be underestimated, but the integrity of the heading (in its role of conveyer of intelligence) must necessarily take precedence over any attempt to structure it so that it files "automatically." Or, to put the matter succinctly, headings do not exist merely to be filed.

Rather thus eschews revision of headings for the sake of filing convenience, but he also advocates abandonment of the "as if" or "file as spoken" tradition, and for essentially the same reasons.

> Variations in form among name and subject headings are an essential part of the structure of a file arrangement; they should not be ignored in filing. It is illogical to construct a heading on way and then to file it as if it were constructed another way.

The three basic principles of his code are

— Elements in a heading should be taken in exactly the form and order in which they appear.

— Related entries should be kept together if they would be difficult to find when a user did not know their precise from.

— A standard set of fields should be established for each major type of filing entry.

The first principle "emphasizes the way a heading looks, not how it sounds," the second "preserves important values offered by the present arrangement of the LC catalogues," and the third "applies the legal precept 'De minimis non curat lex' (the law cannot take care of trifles) as a way of preventing the proliferation of special rules."

Since Rather's rules have already met with general acceptance at LC and been used to produce several book catalogs, and since a computer program already exists to arrange MARC entries following these rules, their influence is likely to be evident in cataloguing practice for some time to come.

On-line techniques

Another important influence on the arrangement of cataloguing information has been the use of on-line techniques. By 1968, Stanford University's on-line BALLOTS project had evolved a technique of filing machine-readable catalogue entries not by any of the usual rules, but in an arbitrarily arranged file with indexes by various bibliographic elements, such as author, title, topical subjects, and so forth. By itself, this was not new. The significant and important aspect of the technique, however, was that the position of each entry in the author and title indexes was computed directly from the data itself, using the first three letters from the author's last name and the words of the title.

The computer, of course, followed the same procedure in retrieving the index entries on command. In effect, the actual filing arrangement of the "catalogue" was transparent to the user. This technique of "derived keys" was being studied by several researchers at about this time. Nugent described four different "compression coding" techniques, and Kilgour and Ruecking reported the results of separate tests using various combinations of characters from author and title entries for retrieval purposes.

In 1970, Kilgour and two associates analyzed the effectiveness of derived keys varying from four to eight characters in length for retrieval of name and title entries taken from a sample of 132,808 entries. They concluded that a "3,3" key—that is, the first three characters of the author's last name and the first three character of the title-would retrieve five or fewer entries 99 percent of the time. Further experiments with titles only (using the first three letters of the first two words, or the first three characters of the following three words) produced similar results.

The results of this research led to the use of 3,3 keys for name and title searches and 3.1,1,1 key for title-only searches in the design of the Ohio College Library Center (OCLC) system, and the technique has been one of the major factors in OCLC's success. The implications of this practice, while perhaps not immediately obvious, are nevertheless enormous. In effect, when the user names a key of six characters, he designates what Kilgour has called a "microcatalog" of five or fewer entries, on average, or twenty or fewer entries 95 percent of the time.

Because ten entries can be displayed on the screen of the on-line terminal used by OCLC participants, and twenty entries require only two screen displays, the desired entry is easily chosen. "Such a catalog," comment Kilgour, "makes it necessary to include bibliographical embellishments required for entries in huge card or book-from catalogs," and of course it also makes unnecessary the solution of the filing rule problems discussed earlier.

Nor are the implications of such on-line techniques necessarily limited to OCLC participants, at least not for long. At a meeting in 1975 of the Association of Research Libraries, Rather reviewed the difficulties of improving existing card files or converting them to machine-readable from, and proposed a "radical solution": "What if the MARC data base were considered to be complete for cataloguing purpose?" This would enable LC to begin on-lie cataloguing on a production basis, and close its existing card catalogs, as several other research libraries already have, or have proposed to do. But it would also have widespread effects on cataloguing practice throughout this country and abroad.

The name and subject authority files used by LC to establish the headings which appear in cards and in its book catalogs, as well as in the MARC tapes, would be limited to names and subjects already on MARC records, and the call numbers would be formed with reference only to a "shelflist" composed of call numbers on MARC records. But, again quoting Rather, "the future would no longer be constrained by the past."

It is characteristic of technical revolutions that we are unaware of them while they are in progress. It is also characteristic that it is not so much the technology that makes the revolution but an idea. Someone must see that the use of a technology in a particular way can change fundamental concept of perception and action. In this sense, Kilgour's concept of "microcatalogs," and the use—now predominantly by library staff members but increasingly by the public—of on-line terminals, may revolutionize the heart of library science, cataloguing.

5

Information Retrieval

Since the 1960s, and particularly since the mid-1970s, two other public services functions have become increasingly automated. One is in the provision of access to subject-oriented indexes, abstracts, and similar reference of access to subject-oriented indexes, abstracts, and similar reference sources. Through use of such services as DIALOG Information Services, System Department Corporation (SDC), and Bibliographic Retrieval Service (BRS) many libraries have incorporate online database searching into the routine complement of services available to users.

As a reference function in libraries, searching and retrieval from machine-readable database offered though these services is only about a ten-year-old phenomenon. The other public service function to be increasingly automated has been interlibrary lending. Both database searching and interlibrary being are library service that predate computers. Automated database searching has a direct precedent in subscription to printed indexes and abstracts covering general or specialized bodies of literature. As for interlibrary lending, libraries relied on printed catalogs, telephones, mail and assorted other means of document delivery long before the vent of the computer.

There is a logical continuum between these two services, both in their noncomputerized and their computer-assisted forms. Indexes and abstracts are likely to provide the library user with references to materials that are not available on the immediate premises, and it is at this point where an interlibrary loan service takes over. In a constantly increasing number of libraries, subject-oriented printed indexes and abstracts, such as Chemical Abstracts, Sociological Abstracts, and Modern Language Association Bibliography, are being replaced or supplemented by online access to machine-readable versions of the same publication.

Use of these database is through a service agency that provides access to many such database. Simply by dialing a telephone number, a terminal in a library in connected via one of several telecommunications networks to the service agency's computer. Through a simple identification, the searcher specifies which database on the computer is to be searched and then proceeds through interactive dialog to retrieve citations that are pertinent to a user's search request. Machine-readable database that are accessible in the manner described above and the computers on which they are stored are so numerous, so constantly changing, and so undocumented that they defy estimation.

Therefore, discussion is limited here to services that have targeted libraries as a primary commercial concern. Even so, it is possible to arrive at meaningful estimates only by setting up fairly strict criteria for inclusion. Martha Williams, for example, estimate that by the end of the 1970s there were 650 "bibliographic-textual-natural language" data-bases searchable either online or in batch mode and "publicly available" on "major U.S. and Canadian bibliographic retrieval systems".

These parameters make it somewhat easier to identify the major bibliographic retrieval systems and services within a library context. In the past, very few actual producers of these databases-primarily the indexing and abstracting services-have felt it economically viable to offer online access themselves. Instead, thee producers have generally concentrated efforts on just creating machine-readable files of bibliographic information, the "raw material" for a retrieval system.

The producer then license the machine-readable database to a database vendor which has developed a retrieval system and has mounted many such databases on its computers. It is the vendor, then that provides libraries and other institutions online access to the content of the databases. Since most of the major vendors have this type of arrangement with a number of database producers, there are fewer database vendors than producers. In some, cases, the relationship between the information retrieval service the retrieval system it uses, and the database to which it offers accesses is fairly straight forward.

For example. System Development Corporation (SDC) offers a service through its internally developed OR-BIT retrieval system, with access to more than 70 distinct database. In other case, the relationship is less direct. For example, Bibliographic Retrieval Services (BRS), another major online service uses a version of the STAIRS system developed by IBM. Perhaps the most complex relationship, especially in a historical sense, concerns the National Library of Medicine (NLM), which acts both as a database producer and as a vendor of its own databases.

The retrieval system used in NLM's own service is ELHILL-2; the earliest version was developed by SDC based in large part on ORBIT. Adding to all this complexity is the fact that a number of the databases available from major vendors are not offered by only one service, but rather by two or three or more. For example, access to portions of NLM's most comprehensive database, MEDLARS. is offered not only by NLM directly but also through BRS and DIAIOG. The ERIC database complied by the National Institute of Education is offered through BRS, DIALOG, and SDC.

As one might expect in such cases of parallel service, there is a significant amount of professional literature that compares the retrieval effectiveness of professional literature that compares the retrieval effectiveness of different service for database offered through more than one service. Two of the practical advantages that online indexing and abstracting databases have over their printed counterparts are the same is in Online technical processing systems and publications catalogue currency of information and flexibility in searching.

In terms of implementation, the earliest of the online technical processing systems, and even online public-access catalogs, predated use of formally established information retrieval services by libraries. This is not to say, however that the design of the latter was in any based on the former. The fact that the information retrieval services did not really emerge until the early 1970s in a publicly available forum was not clue to a lack of developed systems, but rather to other factors. The various information retrieval services had several features in common that distinguished them from online library catalogues and systems for technical processing application. Some of these distinguishing features included:

— Subject orientation, covering material pertinent to a particular discipline, in some cases narrowly defined and in others quite broadly defined, rather than the holdings of a particular library.

— An emphasis on indexing and abstracting of the periodical literature rather than books.

— An emphasis on remote access to centrally stored database by a number of different users.

— Use limited to searching and retrieval of information; data in the master file could not be permanently added to or purged from the database by remote access users.

In addition, the vendor-based information retrieved systems of the 1970s and their predecessors in the 1960s shared a distinguishing emphasis in the direction of system capabilities. Almost without exception, the earliest online library operations systems of the 1960s were very rudimentary in certain features, such as depth of indexing of machine-stored records and breadth of retrieval capabilities for accessing those records.

In contrast, bibliographic storage and retrieval systems developed by the information retrieval services had placed their most concentrated focus on precisely those features so lacking in sophistication in the early online library systems. They are designed with primary emphasis on in-depth indexing of records and exploited the tremendous power of computers in order to provide more sophisticated avenues of retrieval. One of the more successful initial efforts in the design of an online bibliographic retrieval system was in conjunction with the Technical Information Project (TIP) at the Massachusetts Institute of Technology in the early 1960s. The database created for the field of physics; by 1966, the database included about 35,000 citations.

The input for each source article consisted of a citation that included the name of the journal, the issue and pagination on which it appeared, the author and title, the institutional affiliation of the author, and a list of all references cited in the article. The procedure for retrieving citations from the TIP database required the searcher first to specify the parameters of the search, such as which issues of which journal were to be searched. Next, the user entered a search strategy, instructing the system to retrieve all citations of articles, for example by a given author or with appertain term in the title.

Boolean searching could be performed and the system also allowed for retrieval on truncated search terms. In entering more than one term in a title search a user had the flexibility of specifying whether these terms had to appeal an anywhere within a title or whether they had to appear exactly in the same order as specified in the search statement. The decision not to assign additional subject descriptors for each article was based in part on the premise that if very sophisticated search and retrieval capabilities were designed into the system, subject-access avenues might be available that would not depend on manual selection and assignment of subject terms.

Such an experiment had profound implications for creating information retrieval database, for it reduced the input procedure to a clerical task not requiring the additional, purely intellectual, time and effort involved in selecting and assigning subject terms. The nontraditional form of subject access in the TIP system was based on manipulation of citations listed as references in source documents. A searcher could enter an author's name on title of a work and retrieve records for all source documents that had cited that author or titles as a reference. One extension of this capability was bibliographic coupling. A searching could enter two authors names or two titles. For example, and would retrieve records for all source documents that had cited both authors or both titles.

The premise for citation retrieval and bibliographic coupling as a form of subject access was that there exists a strong correlation between the subject article and the subject of works cited therein. Further, the more references any two articles have in common, the before closely related the subject of those two articles are likely to be. For a searcher who was aware of important works in a particular field that were likely to be cited, these retrieval capabilities represented a viable form of subject access.

Recognizing that this was only a partial approach and would have assumed that the user was already aware off articles likely to be cited, the TIP retrieval system design incorporated features that automatically assisted the search process. A most interesting strategy was a set of procedures referred to in TIP as Strategy A. The user did not have to execute all of the steps involved, but merely had to initiate the strategy by entering a search term that was likely to appear in titles of source documents on a subject in which he or she was interested.

After retrieving records for all source documents with this term in the title, the system automatically examined all references cited in those documents and ranked the references according the number of documents citing them. Given a cut-off point the system then took the most commonly cited references and searched for other source documents in the database, which even though lacking the original search term in their title, cited these same references.

Once this second file of source documents was complied and added to the first, the system then examined all titles to identify the most frequently appearing terms apart from the original search term. It then went back through the entire database to retrieve still other source documents with these secondary terms in their titles. This interplay continued until no new records were retrieved. In addition to citation-retrieval and bibliographic-coupling techniques such as those designed into the TIP system, other retrieval techniques developed during the 1960s explored the computer's potential for elimination the need for human indexing and at the same time for assisting in making judgment about the relevance of records in the database to any given search query.

In contrast, a number of other experimental systems were designed in which human indexing of source documents played an integral role. The retrieval effectiveness of these systems was evaluated with regard to such variables as depth of indexing, selection of index terms form various portions of the source documents, use of controlled versus uncontrolled indexing vocabulary, and construction of thesauri. The 1960s were indeed an exciting period for exploring a broad range of issues related to bibliographic storage and retrieval on computers.

DIALOG AND ORBIT

The sophisticated techniques devised for the TIP project and for other system during the 1960s provided the theoretical basis that would be used in the early 1970s. Some of the vendors of that time had, in act, been in the forefront of information retrieval systems design during the 1960s. Of the systems developed in those years, the two that have had the most long-standing impact on libraries are the DIALOG system, developed by the Lockheed Corporation, and the ORBIT system, by System Development Corporation (SDC). By 1964, the Information Science Group at Lockheed's Palo Alto, California, Research Laboratory was experimenting with an online information retrieval system called CONVERSE.

In its initial version, CONVERSE was a primitive effort compared with what it would soon lead to. Searching a database of 100 document citations, experimenters selected prepunched cards on which possible search terms had been recorded, and read these "search cards" into the system through a one-card reader. The system searched the database to find records on a teletype machine, showing either full bibliographic information for each citation retrieved or only the document citation reference number, depending on the number of citations retrieved.

If the search retrieved more than 15 citations, the system sent a message to this effect and the searcher had to refine the search strategy. By 1967, through constant refinement and introduction of more so-phisticated capabilities, CONVERSE was transformed into DIALOG, The impetus for the initial application of DIALOG was Lockheed's contract with the National Aeronautic and Space Administration (NASA) to develop a remote retrieval capability for the NASA data-base, to be used by the NASA Ames Research Center in California.

A year earlier, NASA had awarded a contract to the Bunker-Ramo Corporation for development of a prototype system to allow three other NASA centers remote access to a NASA database stored in Bunker-Ramo's computer in New York City. The 300,000 citation NASA database was mounted on a computer a Lockheed's Palo Alto facility. A remote-access terminal was placed at the Ames Research Center, from which engineers and scientists accessed the database through the DIALOG retrieval language. A subsequent contract was awarded to Lockheed to provided similar retrieval system that could be mounted at NASA's own facilities. The resulting system, called RECON (Remote Console), was base largely on DIALOG.

The rights for future development of RECON were retained by NASA, and in the early 1970s, NASA in fact removed its database storage from, NASA awarded the contract for maintenance and further development of the RECON system to Informatics, Inc., a Lock-heed competitor during the early 1970s in the emerging information retrieval services industry. The System Development Corporation had participated in the design of interactive retrieval systems as early as 1960. This experimental retrieval system, called PROTOSYNTHEX, incorporated full-text searching capability with the text of Golden Book Encyclopedia serving as its database. During the mid-1960s. SDC continued to work on the design of online information retrieval systems.

One result was BOLD (Bibliographic On-Line Display) which was tested on a database several thousand bibliographic records and abstracts for technical reports issued by the United

State Documentation Center. Another SDC system during that period was COLEX (Central Information Reference and Control On-Line Experimentation), used by the Foreign Technology Division of the U.S. Air Force Systems Command at Wright-Patterson Air Force Base in Ohio.

ORBIT (On-Line Retrieval of Bibliographic Information Time-shared) emerged at SDC from experiences gained through the design of such system as BOLD and COLEX. An important development in SDC's information retrieval systems history occurred in 1970 when it was awarded a contract by the National Library of Medicine (NLM). The contract called for a portion of NLM's MEDLARS machine- readable database to be loaded on a computer a SDC's facilities in Santa Monica, California, and for SDC to develop a retrieval system and to provide remote access to approximately 30 libraries and hospitals throughout the United States.

The project came to be known as AIM-TWX (Abridged Index Medicus Teletype Writer Exchange Network). Much as Lockheed had done for NASA, SDC developed a retrieval system specifically for use with the MEDLARS database. Given the name ELHILL the system was very similar to SDC's own ORBIT, Beginning in June 1970, the AIM-TWX service enabled any of the participating institutions to dial into SDC's computer over the TWX or standard telephone network and conduct online searches of the NLM database using the ELHILL retrieval language. In 1971, after months of experience with the AIM-TWX project, NLM mounted a portion of the MEDLARS database on its own computer and, using the ELHILL system, established its own online search service called MEDLARS file mounted on its computers and accessible through ORBIT.

In the mid-1870s, however, NLM terminated that arrangement and withdrew its database from SDC. Among information retrieval systems development during the 1960s,

ORBIT and DIALOG eventually had the greatest impact on the broadest range of libraries. But another important system appeared late in the decade that would gain library use-the Mead Data Central system developed by Mead Corporation. Its first applications were in providing online access to the Ohio Bar Automated Research (OBAR) legal database of the Ohio State Association, the Epilepsy Abstracts Retrieval System (EARS) database provided by the National Institute of Neurological Diseases and Stroke, and the machine-readable bibliographic files of the National Technical Information Service.

Mead narrowed its scope during the 1970s, concentrating largely on providing access to LEXIS, an outgrowth of OBAR, and to NEXIS, a news-oriented database. Application of online retrieval systems during the 1960s were generally limited to use in government agencies and by corporations. Probably the first Information retrieval "service" used by libraries other than in those agencies or corporations was not provided by Lockheed or SDC, but the State University of New York Biomedical Communications Network (SUNY BCN). Under the direction of Irwin Pizer SUNY BCN became access via direct telephone lines to a database storied on a computer at the SUNY Upstate Medical Center in New York.

The libraries included seven in New York State, the Country Library of Medicine at Harvard, and the National Library of Medicine. As information retrieval services emerged in the early 1970s, SUNY BCN was somewhat distinctive in that it was neither the major producer of the database used nor the developer of the information retrieval system used to access it. The original retrieval system used was the Document Processing System (DPS) developed by IBM, replaced in 1973 by IBM's STAIRS (Storage and Information Retrieval System). Although it is true that the database was in part created by direct input from member libraries, much of it consisted of MEDLARS records provided by the National Library of Medicine.

The SUNY BCN database was broader in purpose than those eventually accessible through other information retrieval services. By 1969, it included MEDLARS files dating back to 1964, bibliographic records for monographs and serials titles cataloged by NLM since 1967, for all monographs cataloged in three of the participating libraries since 1962, for a portion of monographs cataloged by the Countway Library since 1968, and records for a 25,000-title union list of serials held libraries on campuses of State University of New York and in several others. Not all of this data was permanently available online, but requests for searching the file stored offline could be submitted over the same terminal used for online searching. These requests were then batch searched and the results mailed.

In terms of retrieval capabilities the early version of the IBM Documented Processing System was not as sophisticated as some of the systems being developed elsewhere. Once entered, a search could not be interrupted by the user until it was completely processed. Even more restrictive the system could return only ten citations in response to any search query. If the search retrieved more than ten citations, the system printed out only the ten most recently published.

If the user wished to see more than ten, the search strategy could be reentered and processed again to retrieve ten more citations. This was on awkward and time-consuming procedure, particularly since the terminals communicated at comparatively slow speeds. Many of the design limitations imposed by the IBM Document Processing System were alleviated when SUNY BCN converted to the STAIRS System. SUNY BCN grew from its original 9 participants in 1968 to 32 by 1973 and continued in operation until 1977.

Early Database Online

As is evident from the description of the early information retrieval systems and services that were to have the strongest impact on libraries, the machine-readable MEDLARS database played a key role in their development. By late 1971, the MEDLARS database or portions of it were available though at least three different services: SDC, SUNY BCN, and the National Library of Medicine itself. NLM may have had one of the largest bibliographic machine-readable databases, but other government agencies were also creating such files and mounting them for information retrieval were also creating such files and mounting them cress with the NASA project, was in the forefront of providing government agencies with retrieval services through the DIALOG system.

In 1969, it mounted the Nuclear Science Abstract database for use by the Atomic Energy Commission and the ERIC database containing machine-readable records for citations in Research in Education and Current Index to Journals in Education for use by the U.S. Department of Education. Between 1970 and 1972, other information retrieval services began providing access to database created by government agencies: SDC made the ERIC database available through ORBIT; Mead briefly provided access to the NTIS database; and Informatics, Inc, maintain the TOXICON (later TOXLINE) retrieval system for National Library of Medicine.

Of equal important to the increase in the number of database produced by government agencies was the mounting of nongovernment databases by the information retrieval services. By 1972, SDC was offering online access through ORBIT to the Chemical Abstracts Condensates database; Mead was offering access to the Psychological Abstracts database; and Lockheed was providing access to PANDEX created by the McMillan Company and covering scientific and technical literature.

Packet-Switching Communication N etworks

During the early 1970s, online access to at least some of these data bases was becoming publicly available. In essence, any institution could gain access if it was willing to pay the fees charged by the service on whose computers the database stored. But there were a couple of problems that stood in the way of an immediately outburst of enthusiasm on the part of librarians. One was a basic unfamiliarity with such services, their technology and the procedures involved. Beyond this however, was the question of costs, and most particularly of communications costs.

In accessing a database through one of these service, it was necessary to dial the computer via standard telephone lines or in some cases, through a TWX network. Since the service's central computer might be located at some distance, this was not an inexpensive venture, and communications costs could comprise a substantial portion of the total cost of the search. This obstacle was partially overcome during the early 1970s with the introduction of public accessible packet-switching networks, such as TYMNET and TELENET. Circuit-switched networks, such as used for telephone and TWX communication, reserved a fixed bandwidth over which communication occurs for the entire duration of a connection.

Packet switching, on the other hand, sends messages over communications networks in smaller blocks of information. It ideally suite for communication in a computer network, since the network is activated for any particular user only when a message is ready to be sent. It does not maintain a permanent connection between points over a field bandwidth during the entire course of a communication, which would include time spent in formulating search strategy, keying the strategy, computer processing of a request, and so on.

Extensive research on the cost-effectiveness of packet-switching communications networks had been conducted by Rand as early as 1964 Commissioned by the U.S. Air Force to conduct a feasibility study the Rand Corporation in fact recommended that the Air Force adopt such a network, which would not only be less expensive but would also provide a defensively more secure communication network than was available through existing systems. The Air Force adopt also provide a defensively more secure communication network than was available through existing systems. The Air force did not act on the recommendation.

In the Late 1960s, however, the Advanced Research Projects Agency (ARPA) of the U.S. Department of Defense funded development of a packet-switching network. By December 1969, ARPANET was declared operational, linking four computers located at deferent sites. In the late 1960s and early 1970s work proceeded on establishing similar network for a broader audience of users. These efforts culminate in the formation of the TYMNET and TELENET network the latter created by some of the people who had worked on ARPANET. Each of these "value added" networks established nodes in a number of major U.S. cities.

Each node consisted of a minicomputer, which acted as a communications link and as an interface device between remote terminals and host computers, such as those on which database were being stored at Lockheed and SDC. In order establish a connection between a terminal and a host computer the user still had to dial into the computer but in using TYMNET and TELENET, the standard telephone link occurred only between the terminal and the minicomputer at the nearest network node. Messages received by the minicomputer were then sent over telephone lines leased by the networks to another node, and so on through the network until the message reaches the host computer. Messages returned from a host computer to a remote terminal followed the same procedure.

The exact route any message might take through the networks was determined dynamically by the network itself and depended on the flow of traffic at the precise moment of communication. The bottom line of the packet-switching networks was that they dramatically reduced telecommunication costs. As a measure of the magnitude by which these costs were reduced, Devis McCarn at the National Library of Medicine estimated in 1974 that with the introduction of packet-switching networks, the average cost of communications for MEDLINE searching immediately dropped from $24 per hour to $5 per hour- a decrease of about 80 percent.

A further innovation was the pricing structure. Unlike the telephone system, the packet-switching networks charged a flat hourly rate for communications, regardless of which node a message emanated from and regardless of where it was sent. During the same period that the packet-switching networks were being introduced commercially, computer hardware costs were declining sharply. This was no small consideration for the information retrieval service as they added new database to their collections.

By mid-1974 for example, SDC was offering service on about 11 database with a collective 3.5 million bibliographic records. These require in the vicinity of 3.3 billion bytes of storage on disks that had a capacity of 200 million bytes each 11. These dual-density disk were in fact a relatively recent breakthrough on the market, improving upon disks that had a storage capacity in the range of only half that amount. The decrease in costs for telecommunications and per-unit computer storage brought the price of these service into a more acceptable range for libraries and other institutions.

As the number of users began to grow and maintenance and development cost could be distributed across a large number of users, prices decreased still further. The increased revenues, in turn, could be used to license new databases and make additional enhancements in order to attract still more users. As an indication of the total impact of these and other factors on pricing during the early 1970s, the cost of searching through Lockheed's service in 1973 was less than half of what it had been in 1970, and between and 1973 and 1975 the cost was again reduced by about half.

Although it is difficult to trace the magnitude by which the use of information retrieval services grew during the 1970s, one is to examine the growth of the service themselves. In 1971 there were less than half a dozen services that offered access to bibliographic databases, and each of those had only between one and three database available. Lockheed and SDC each offered service to more than ten databases, and NLM had broadened the number as well, although some were essentially spin-offs from the main MEDLARS file. SUNY BCN continued its service and in fact entered an arrangement with NLM for its computer to act as a backup for searching MEDLARS. One of the new services to appear in the early 1970 was the New York Times Information Bank (NYTIB). Its database had been available for in-house use at the New York Time office since 1971.

In 1973 access became publicly available exclusively through the Times own new commercial information retrieval service, using a retrieve language designed by IBM. Coverage included not only articles from the New York Times dating back to the late 1960s, but also selects items from more than 60 other newspapers and periodicals indexed back to 1971. The depth of the NYTIB indexing was extremely through, based on the use of a controlled vocabulary. The was at time cause for annoyance, especially given the fact that this database was a source of material on current events. One user, for example wrote of the frustration in attempting to retrieve articles about migration to the "sunbelt states" at a time of extensive news coverage of that phenomenon.

The controlled vocabulary used by the NYTIB did not include such a term, nor did it have any term that corresponded to the geographic area being described. Although the descriptor "Southern Rim State (LS)" was eventually added to the list of acceptable index terms and an attempt made to reindex some of the material in the database retrospectively with this term, such efforts required time. This aspect off a controlled vocabulary was certainly not unique to the NYTIB; it is one faced by any database using a controlled vocabulary for searching. Martha Williams estimated that in 1974 approximately 700,000 online searches were conducted in the United States on remote-access bibliographic retrieval systems other than OCLC.

To illustrate how pivotal a period this was for the commercial information retrieval service, in December 1972, SDC could claim less than one-half dozen organizations using its searching service. Within six months, this number had grown to more than 40, and by early 1974 the number of subscribers approaches 150. Another author stated that the number of searches on Lockheed's DIALOG in 1975 was more than seven times the number conducted in 1973. Already by 1974, there sere more than 500 organizations searching MEDLARS through either NLM or SUNY BCN services, and Cuadra estimates that in 1975, 18 well over 2,000" organisations were conducting searches through the various on line information retrieval services. With such a rapid growth rate, there were constant changes and new problem.

One very basic problem for the service themselves has how, despite advances in technology, to keep their computer configurations of sufficient size to handle the enormous number of machine-readable records included in the databases to which they offered access. One solution was to maintain records online only for the most recently published materials, often a two-to five-year span. SDC used an approach whereby database were rotated during the course of each day, making some available for the entire day but others for only a few hours, a decision that resulted in a one-half to two-thirds savings in storage costs over what would have been necessary had all databases been mounted all the time the system was available.

6

Interlibrary Lending

Beyond the initial most basic step of identifying what is to be borrowed-which libraries assist by providing bibliographies, printed abstracting and indexing journals, and more recently online information retrieval services-interlibrary lending is a process involving several steps. The steps that involve the greatest degree of coparticipation by libraries and group of libraries: distribution of locational information; communication of interlibrary loan requests and related correspondence; and physical delivery of materials form one library to another. It excludes, for the most part, steps that are important in the overall process but are primarily internal to each of the participants; retrieval of materials from bookstacks at the lending library, wrapping and preparing materials for shipment, bookkeeping function, and so on.

Before the 1950s, the tools used in communicating interlibrary loan requests, and delivering materials changed very little. Communication was by male or telephone, and delivers was generally by mail with bus or automobile courier service sometimes playing a role. The area of greatest progress during the first half of the Twentieth century involved refining the ways to determine the location of an item to be borrowed. The most pertinent developments included union lists, surveys of resources, union catalogue, and bibliographic centers.

Although these development were not a result of automation. They are a necessary back ground to the events that were to follow after 1950. The most specific aids for locating materials were union lists and union catalogues. A popular form during the first half of the twentieth century consisted of compilations of periodicals or serials holdings. Many of these compilations were published and distributed, although a significant number were maintained only in card form and stored at one of the participating institutions. Possibly the first printed union list of serials was the List of Serials in Chicago, published in 1901.

If that was indeed the first, it was followed not long after by several others, and in subsequent decades union lists of periodicals and serial were complied in a great many locations. The late 1930s was a periods of intense activity for union lists of serials and periodicals, as well as for union catalogues of broader scope. Ralph Esterquest noted that in examining the cumulation of Library Literature for 1936 through 1942, one could find reference to union listing projects in Atlanta, Boston, Denver, Maine, New York, Ohio, Philadelphia, San Francisco, southern California. Tennessee, and the southern United States in general, and these represented only projects being reported in the professional literature.

Although there was considerable activity in compiling union lists at the local level, there were also significant projects conducted on a broader regional and even national scale. Prominent was the publication by the Library of Congress in 1926 of the Union List of Serials containing 75,000 titles and representing the holdings of 225 libraries. A second edition was published in 1943, containing 115,000 titles and representing the holdings of 650 libraries. At the local level the most elementary form of a union catalogue containing more than periodicals or serials holding consisted of the exchange of catalogue cards between two or more libraries.

This also occurred beyond local projects, exemplified at the most comprehensive level by the National Union catalogue (NUC) assembled by and house at the Library of Congress. It was started in 1901 as a main entry union card catalogue, and its contributors included government libraries in the Washington, D.C., area, the Library of Congress itself, and several prominent library throughout the United States, including the New York and Boston public libraries, Harvard, the University of Illinois, and the John Crerar Library in Chicago. Additional libraries joined the effort, and by 1927 the catalogue contained nearly 2 million holdings. By 1954, it contained cards for nearly 7 million titles and 13 million holdings contribute by more than 500 libraries.

In the mid-1950s, the first monthly, quarterly, annual and five-year cumulative printed editions of the National Union catalogue were published in book forth. This still left the question of what to do with the contributions of the previous 50 years, which remained available only in card form at the Library of Congress and various NUC card depository libraries around the United States, where duplicate catalogue were maintained. A contract for publishing the earlier holdings was eventually awarded to Mansell, Ltd., and in 1989 the first volumes of the National Union catalogue, Pre-1956 Imprints appeared. More than a decade later, the last volume of the more than 700-volume set was printed and distributed. Carried our largely during a time of increasing reliance on computers for bibliographic and locational information, the publication of the Pre-1956 Imprints remains a landmark in library cooperation.

Of the major tools maintained or publish during the first Half of the twentieth century to assist libraries in locating materials for borrowing through interlibrary loan, a less item-specific type than either the union list or union catalogue was the directory of resources. Such directories seldom attempted to identify specify titles held by libraries, but libraries, but rather described the general nature of various library collections, particularly noting areas of emphasis and strength. One early work was William C. Lane's and Charles K. Bolton's Notes on Special Collection in American Libraries published in 1892.

Descriptive guides to resources published during the first several decades of the twentieth century, like union catalogue and union lists, varied a great deal in scope and content. Some focused on resources in a particular type of library: either within a limited geographic locale, such as resources in special libraries in an urban area; on a broader region, such as the Special Libraries Association's four-volume Special Library Resources, published between 1941 and 1947, Other directories of resources placed their primary focus on geographic parameters and encompassed all or most types of libraries within those bounds, ranging from city, such as Robert Down's Resources of New York City Libraries published in 1942, to a broad region, such as the Resources of Southern Libraries, 1938, also prepared by Downs.

The purpose of these directories extended beyond their application in interlibrary loan, but given the state of title-specific bibliographic control at the time, they often provided general clues as to where materials might most likely be found if specific locations could not be determined from other sources.

The lending of books between libraries is only one part of a broader concept of library cooperation. The tools and institutions described above-union lists, union catalogue, bibliographic centers, and surveys of resources-were the most important aids in determining location during the first half of the twentieth century. But there were other aspects of library cooperation that also had an impact on the way interlibrary loan was conducted or on the level to which it was used. Such developments included cooperative acquisitions projects and cooperative collection development programs.

Some of the larger-scale activities led to the founding of special interlibrary collection centers during the 1940s and early 1950s. Some of these centers, such as the New England Deposit Library, were primarily cooperative storage areas, with little attempt by participating libraries to organize the materials into an integrated collection. Others, such as the Hampshire Inter-Library Center (HILC) and the Midwest Inter-Library (MILC) started off mainly as storage centers but development into something more. The materials stored by members were merged into a collection and duplicates were often purged. Important materials that members did not wish to acquire independently were published collectively, stored at the center, and borrowed from there through interlibrary loan.

Of all the centers established in the 1940s and early 1950s, the most successful has been the Midwest Center. Established in Chicago in 1951 by ten midwestern research libraries, the center changed its name in 1965 to the Center for Research Libraries (CRL) reflecting its broadened geographic basis beyond the midwest, and has continued to operate and expand its scope of activities.

Two important technological innovations introduced into libraries before the 1950s. which had implications for interlibrary lending, were microfilm and photocopying. Microfilming and photocopying could potentially easy a lending library's conflict between the desire to cooperate with other libraries and its primary allegiance to its own clientele. By sending a microfilm or photocopy of a document rather than the document itself, the lending library could fill an interlibrary loan request without having to remove the original document from the immediate premises for an extended period.

Photocopying emerged as the preferred and more frequently used technique and eventually become the primary method for providing journal articles and briefer documents requested through interlibrary lending. With the exception of microfilming and photocopying the tools to facilitate interlibrary lending before 1950 were aimed almost exclusively at one aspect of the process-determining the physical location of materials.

Until the 1960s. and actually until the early to mid-1970s there was little change in technology even in these tools. Libraries continued to rely on printed sources—surveys of resources, union lists and cards in union catalogue—as their primary sources for locating items. Although such sources may have grown in number and scope through the 1950s and early 1960s, the methods of producing them did not. Using tools created in the first half-century, libraries in the 1950s began to look at another aspect of the interlibrary loan process-communication technology.

For the next decade and a half, this attention would focus primarily on the adoption of teletype machines as a way off transmitting interlibrary loan requests and associated administrative correspondence. The use of Teletype, or TWX (TeletypeWriter Exchange), significantly reduced the time required for communication where compared with mail services. Although the telephone was as speedy as TWX, it required another person's immediate attention at the receiving end; TWX did not. Furthermore, a TWX transmission resulted in a printed request at the receiving end, while a verbal request required the receptor to jot down details.

Using TWX, communications time, and there-force on the machine. TWX machines equipped with a paper-tape device allowed an operator to type a message into a machine-readable paper tape in the local, noncommunicating mode. After recording the full message on paper tape, the TWX number of the receiving library was dialed. Once a connection was established, the paper tape was fed through automatically and its message transmitted at 60 to 100 words per minute. Teletype technology had been available long before its more widespread adoption in libraries.

As early as 1927, the Free Library of Philadelphia installed teletype machines for internal communication between the circulation desk and the bookstracks, but it was not really until the early 1950s that libraries began to adopt the technology for communicating with each other. As with any alternative communications technology, there was some reluctance to adopt TWX since its usefulness to libraries would be realise only when there was an ample number using the same technology. As a result, many of the instances in which TWX was adopted during the 1950s involved groups of libraries that had already developed, or were developing an established and predictably high level of communication with each other.

One of the earliest examples of TWX use for interlibrary communication was between the Milwaukee and Racine public libraries in Wisconsin, dating back to 1949. With a high volume of interlibrary lending, the two libraries felt TWX to be an economical alternative to the telephone. On a broader scale, the Midwest Inter-Library Center in stalled TWX in 1951 as an alternative means 16 members of the center had installed TWX in libraries.

Also in the early 1950s, TWX machines were installed in the Michigan State Library, the Detroit and Grand Rapids public libraries, and the University of Michigan for rapid transmission of interlibrary loan requests and other messages among the four. About the same time, libraries on the eight campuses of the University of California installed TWX for interlibrary communication. By the mid-1950s, the costs associated with TWX equipment had escalated considerably over what they had been just a few years before.

Several libraries that had installed TWX during the early 1950s decided that the level of fuse did not justify the continued rental of the machines, and had them removed. Michigan did so, as did some members of the Midwest Inter-Library Center, at least on a temporary basis. Infact, in a 1958 survey. James Mack could account for only 24 libraries with TWX installations. Probably an underestimation of the total number of libraries inn the United States using TWX for communications, but nonetheless indicative that TWX was hardly a common means of inter library communication at that time. The use of TWX in Libraries increased during the early 1960s.

In 1964, a directory of TWX subscribers listed 64 libraries has having their own equipment. By the following years, the same publication listed 130 libraries. This dramatic increase can be accounted for by the emergence of several not working situations that adopted TWX in 1967 as their primary means of communications. Included among these were the National Library of Medicine and several medical libraries along the Atlantic coast and in the Southeast, and a few state networks such as in Indiana, where 25 libraries installed TWX in 1965.

Facsimile Technology

Although teletype equipment was being installed in a gradually increasing number of libraries between the early 1950s and the mid 1960s, its use in interlibrary lending was necessarily restricted to the transmission of requests and associated brief correspondence rather than of the materials themselves. Both at the beginning and end of this period, some libraries experimented with electronically transmitting the actual text of the materials requested. These efforts, which focused on the use of telefacsimile equipment, were few in number generally experimental and in most cases short-lived.

The first major experiment with facsimile transmission of documents from one library to another was carried out by the Oak Ridge National Laboratory in Tennessee in 1951. The System developed by RCA, consisted of two primary pieces of equipment. In the parlance of the day, one was a "reader-transmitter "located at the central library, and the other a "recorder" located at a test site eight miles away. When activated, the reader-transmitter photo-graphically scanned a page of printed matter and converted the impulses of reflected light into electrical signals. These signals were then transmitted over a normal telephone line to the recorder equipment at the receiving site.

When received, the signal were converted through an electrolytic process back into appropriate images on a sheet of paper, forming a facsimile of the original document. From the time of placing a page of a document on the reader-transmitter, it took 4.5 minutes for the facsimile copy to be complete at the receiving end. In retrospect, one of the most interesting aspects of the Oak Ridge experiment was that the reader-transmitter was capable of copying from a volume, a capability that is still not available on any telefascimile equipment commercially produced in the United States today. It was not until the mid-1960s that libraries began to re-examine the implications of telefacsimile technology for interlibrary lending.

The telefascimile technology of the mid-1960s operated on the same basic principle described earlier, but was greatly improved over what it had been in the early 1950s. In 1966 and 1967, three important experiments and programs were initiated by libraries. Two were funded by the council on Library Resources (CLR) and involved academic libraries in the western United States, and the other was carried out in connection with state networking activities in New York. The first of the CLR-funded experiments, in 1966, was based at the University of Nevada at Reno (UNR) and involved facsimile transmission of documents from the UNR library to the campuses of the University of Nevada at Las Vegas and the University of California, Davis.

The second CLR-founded study, in 1967, involved facsimile transmission of documents from the University of Califôrnia, Berkeley, to the University of California, Davis. The equipment used in these studies consisted of different models of devices manufactured by Xerox, and the type of communications channel was also deferent: the earlier experiment transmitted over normal telephone lines, and the latter over a dedicated telephone line. Although the results of these two experiments differed in detail, some similar general conclusions were reached. In the 1967 study, Shieber estimated that if personnel activities were better organised the cost of document delivery via telefacsimile would be about $1 per page inclusive of equipment, staff, and communication charge.

In the 1966 UNR study, Morehouse estimated a cost of about $0.685 excluding communications, which could range from hardly anything to as high as $3 per page depending on distance an whether communications were via normal telephone line service, or leased line. Both reports stressed the circumstantial nature of costs given the short duration off the experiments, and pointed out that in log-term practice, costs would depend a great deal on factor such as volume of activity and type of communications link. There were incidences of equipment failures in both experiments, and the quality of reproduction at the receiving end of the transmission was not always satisfactory, particularly for smaller-point type, such as that use din footnotes, charts, and table. Schieber reported that in the 1967 study 14 of 65 respondents to user questionnaires expressed dissatisfaction with the quality of reproduction.

An interesting focus of one part of the 1967 study was on the distribution of time for various tasks during an interlibrary loan transaction. Although a copy of most article-length documents could be transmitted within an hour, delays in processing meant that the average time between receipt of request at Berkeley and receipt of a copy of the item at Davis was more than 6 hours. The average time between receipt of a copy at Davis and the sending of a notification to the borrower was 3 hours. These figures were based on a 9-hours workday and 5-day week, on a stricter scale, taking account of a 24-Hour day and weekends, the time required for completely.

The clear implication is that, even with rapid transmission, the other tasks related to the handling of an interlibrary loan request interjected substantial delay in overall delivers performance. It is also interesting to note that for the same requests, the average time between a patron's notification of receipt of material at the library and actual pickup was 8 hours. This was on a 9-hour, 5-day-per-week schedule; in actual time calculated on a 24-hour, 7-day-per-week basis, the average pickup time was nearly 62 hours after notification.

Between January 1967 and March 1968, the New York State Library incorporated a facsimile transmission program. FACTS, as part of its statewide interlibrary loan network. It was broader in scope than the CLR studies, not only in duration and magnitude but also by the fact that a request for a facsimile document could be submitted from virtually any library in the state, of course, had such equipment. Not all libraries in the state, of course had such equipment, and thus the geographic distribution of FACTS receiving sites relative to libraries where requests originated imposed some limitations on the system Nevertheless, the FACTS program operated for 14 months and transmitted facsimile copies of nearly, 4,000 documents requested through in-terlibrary loan. The program ended in 1968 for several reasons, some of which echoed the conclusion of the CLR studies.

One of the two equipment models was mechanically unreliable and the other was prone to faulty transmission. Of the patrons responding to postcard questionnaires, 22 percent suggested that the quality of copy be improved, almost the same proportion that complained about copy quality in the 1967 CLR experiment. The complex operation of the statewide system though which requests had to be channeled, combined with routing processing time and processing backlogs, resulted in an average time lapse of several working days between the time a patron submitted a request and the time that material was available for pickup.

An interesting aspect of this time lag was that 70 percent of the users responding to a question on the postcard survey indicated that the FACTS program would be useful even if it took two or three days longer to receive the requested material. Although the two CLR-funded studies and the FACTS program were the most important, other programs based on facsimile transmission among libraries was occurring around the same time. By 1966, the State Library of Hawaii was transmitting copies of documents via facsimile to each of three regional libraries located on the outer islands.

Unfortunately, a detailed evaluation off that program was apparently never account in 1966 very soon after it had been implemented. In the same year, several university libraries in Pennsylvania installed telefacsimile equipment. It use was continued until 1975 but essentially it served for most of that time as an alternative means for communicating interlibrary loan requests and other brief correspondence rather than as a system for transmitting the full text of requested materials. Since the early 1950s, many of the libraries installing TWX equipment for interlibrary communication had done so in a relatively informal away, enhancing an already existing pattern of relationship with neighboring libraries or with libraries connected by some other common bond, such as membership in an interlibrary center.

In the mid-1960s, this same technology began to be adopted by large numbers of libraries in a quite different context, one of considerable structure and formality. This context was the state-based library network. Within most states, information ties had developed between groups of libraries over the year, but these patterns were seldom organized into a coherent whole. In 1956, the United States Congress passed the Library Service Act, providing federal funds to assist development of public library services in rural areas. The money was allocated on a state-by-state basis to be administered thorough a central agency in each state, usually the state library.

Although the Library Service Act did not directly establish the basis for networking that emerged a decade later, it laid a crucial foundation by establishing the need for statewide organization to plan and administer federal funds, and by recognizing state libraries as the agencies to serve in this capacity. During the next several years, the Library Services Act underwent a variety of changes broadening its scope, and in 1964 its name was change to the Library Several and Construction Act (LSCA). Although the act had provisions to support interlibrary cooperation as a legitimate end for assignment of funds, Congress added more specific legislation in 1966 as Title III, providing for funding for "the establishment and operation of systems or networks of libraries, including state, school college and university, public and special libraries and special information centers...."

During the late 1960s, a number of states create TWX-based networks, whose major function was the routing of interlibrary loan requests. It would be a mistake to attribute these communication networks solely to federal funding under LSCA Title III. In the first place, until 1970, the federal LSCA monies required a matching amount from the states. Second some states had already laid the organizational groundwork for state networking, although only public libraries were usually involved.

What the Title III funding and its matching state allocations provided in most instances was the financial force to accelerate these plans and to broaden their scope.

The state based interlibrary loan network were established in response to the ever-increasing demand for inter library loan services and were structured in such a way as to address some of the more pressing problems characteristic of interlibrary lending patterns during the preceding years. Apart from scattered local union lists and catalogue there were only a few major sources, such as the National Union catalogue, there were only a few major state or regional union catalogue, on which libraries had been able to rely or title-specific to these major sources were relatively few in number when viewed in a large context and they were generally the largest libraries, which had, indeed, made their holdings known through these sources. This imbalance was compounded by the fact that many librarians who did not have access to any of the major locational sources, sent their requests to libraries they thought most likely to own an item, which usually meant the same large libraries that were also being inundated with known-location requests.

The most direct measure for alleviating these imbalances would have been the compilation, publication, and widespread distribution of more broadly based statewide union catalogue, but in the mid and late 1960s such a solution was not yet practical. In lieu of this interlibrary loan communications networks were established and designed to alleviate at least some portion of the imbalances by providing a structure to redirect the pattern of interlibrary loan traffic among libraries in the state. The structures usually entailed a three-or four-tier hierarchy through which interlibrary loan requests were channeled, with libraries at upper levels usually using TWX equipment to communicate with one another.

A local library either telephoned or mailed an interlibrary loan request to a regional stem headquarters, often housed at the largest public Library in that area. If the request could not be filled by that library it was sent via TWX to a library in the next level of the hierarchy, usually one of several designate major resource libraries participating in the network. In some states, there were one or two steps from flow of traffic, thereby filtering out a portion of the request's that would have otherwise been sent directly to them. Although this was the case in terms of proportions, its impact was often countered by significant growth in the overall volume of activity was fostered by a formal network and by changing attitudes toward interlibrary loan.

The large libraries were still receiving as many requests as they had previously. Many states used Title III and other federal and state funding to establish a reimbursement schedule for lending, thereby alleviating at least part of the financial burden of interlibrary lending operations at the large libraries. One of the earlier state-base network was the New York State Interlibrary Loan Network (NYSILL). Legislative groundwork for networking in New York State began a early as 1950 and in 1958 a library systems law was passed, backed by a stronger program of state aid.

In 1966, NYSILL was formally established as an interlibrary loan communications network funded with part of $700.000 appropriation by the state legislature. The network included 22 regional systems, 3 geographic referral libraries and 9 subject referral libraries, with the state library acting as both a resource center and as a switching center between regional systems and referral centers. When a library in the state other one of the above wished to borrow something, it telephoned or mailed the request to the facility that served as its regional system library.

The system library tried to ill the request from its won collection, but if unable to do so, it transmitted the request via TWX to the New York State Library. If the state library was unable to fill the request from its collection, it transmitted the request via TWX to one of the tree geographic referral libraries, determined on the basis of the region in which the library originating the request was locate. If the request could not be filled by the geographic referral library, it was transmitted back to the state library, and from there to one of the nine subject referral libraries determined on the basis of the subject content of the item requested.

If the request could not be filled at the most likely subject referral library, it was transmitted back to the state library. Some request that went through this cycle without being filled were then transmitted to yet another of the subject referral libraries, but in most cases they were sent back down the hierarchy as unfilled. When any library along the hierarchy could fill a request the material was mailed directly to the originating library. As described earlier in this chapter, copies of documents for some requests filled through NYSILL in 1967 and early 1968 were delivered via facsimile transmission through the FACTS program.

The interlibrary loan TWX networks established in other state followed patterns similar to NYSILL with local variations. In Million, of example the state library was not a separate levels of the hierarchy as in New York but rather served as one of the four research and reference centers, with requests transmitted from any of the regional systems in the state to any of the four research and reference centers. In states where the state library did not have a strong collection of its own, the largest public library in the state frequently assumed the role of switching center.

In state where there were large union catalogue such as in Pennsylvania, Colorado, and Washington, the bibliographic centers housing these catalogue often or as a bibliographic resource center for requests that had not been filled by the time they reached a specified level within the state hierarchy. One of the last-tiered networks established during the late 1960s and early 1970s was the Minnesota Interlibrary Teletype Exchange (MINITEX). Instead of several levels of communications hierarchy and a multiplicity of referral centers at the highest level, libraries with TWX machines in Minnesota communicated directly with a single center, the MINITEX office headquarters at the University of Minnesota drew upon the collections of the university and surrounding libraries in the Minneapolis-St. Paul vicinity to fill requests.

Delivery to other locations in the state including 18 regional public libraries, had installed TWX equipment. The MINITEX fill rate in 1973 was around 85 percent and in three-fourths of these cases, the materials lent or copied had been delivered to the requesting library and were awaiting patron pickup within three working days after the request had been received at MINITEX.

There were a few other state interlibrary loan communications networks that conformed neither to the hierarchical type as exemplified by NYSILL nor to the more centralized structure such as MINITEX The Lowa Interlibrary Teletype Exchange (I-LITE), for example, consisted of three separate round robin teletype wheels, one for eight public libraries and system centres, another for ten smaller academic libraries, and another for four major university libraries. A request was entered on the appropriate type-of-library wheel and circulated with other requests from one library to the next on a daily round robin until it had been filled or had circulated to all libraries without being filled by any.

A request unfilled on its won type-of-library wheel could be placed on one of the other wheels the following day. The state library acted as a switching center among the three wheels. Beginning in the mid-1960s, the number of states implementing interlibrary lending communication networks steadily increased. By 1970 there were 17 states with formal TWX communication networks, By 1980 there were about 40 states with interlibrary loan communication networks. Some had replaced, or were about to replace, their TWX machines with other types of equipment, teleprinters being among the most common.

In 1982, Montana adopted microcomputers as communications devices for its state network, using a round robin setup similar to Low's I-LITE, but simplifying matters by having only one wheel rather that three. During the 1970s, at least a dozen state libraries experimented with facsimile transmission as a possible means for communicating interlibrary loan requests and other brief correspondence to libraries in the state network, but these tests proved un-successful in all but about three states. In nearly every instance where a formal interlibrary loan communications structure was established, there were significant changes in in-terlibrary lending in the state with respect to volume of activity, pattern of traffic, and extent of participation.

Although they made an enormous impact, these networks by no means ever embraced all interlibrary loan activity among libraries in their respective states. The volume of traffic outside the formal state structures remained high and continued to grow as the general level of interlibrary lending activity increased on the broader scale. Some libraries viewed the network structure in their state as designed to actually discourage participation. Even in multitype state networks, there was a common tendency among academic and special libraries to see the networks as emphasizing primarily public and school libraries.

Historically, the role of most state libraries had been in public library development. Combined with this precedent was a cooperative tradition in American libraryship that tended to focus on single-type library cooperation. In many case, the reluctance of academic and special libraries to participate enthusiastically in state networks may have been based more on the psychology of these factors than on any direct evidence of discriminating practices or polices. In other cases this psychology was reinforced by statutory policies. That actually prevented some type of libraries from participating and allowed participation for others only on a restricted or specially designated basis.

In Illinois, for example, certain categories of academic libraries could participate only as associate members of systems for many years priority handling of their interlibrary loan requests at the system level. The type of library most commonly prevented statutory (or interpretively) from any type of participation in state networks was the corporate library. A more common constraining in many state networks, particularly during the late 1060s, was the definition of what types of material could or could not be requested. For example, materials for which requests could not be sent through NYSILL at that time included multivolume sets, in print paperbacks, and current publications in general. Requests could not be sent through NYSILL at that time included multivolume sets, in-print paperbacks, and current publication in general.

Requests for fiction, textbooks, children's books and a broad category of nonsearch materials including self-help books, cookbooks, and books on sports, recreation, and hobbies could be sent up the hierarchy to the state library but would not be sent beyond be sent up the hierarchy to the library, but would not be beyond there to any geographic or subject referral libraries unless accompanied by a statement that the materials were to be used not be used for serious research. Such restrictions conformed to stringent interlibrary nor to the desires of local libraries to satisfy those demands.

A study by Nelson Associates of the early operation of NYSILL show hove great this problem could be. Of 46,000 requests received by the New York State Library during a six months period in 1967, nearly 15,000 could neither be filled by the state library nor were eligible for referral of a geographic or subject resource center. This meant that each months an average of nearly 2,500 un-filled requests for materials that fell into the nonreferral categories described earlier were sent back down the hierarchy from the state library as unfilled requests. Having exhausted the formal network channels, further attempts by the local or system library to fill these requests required going outside the channels.

Perhaps the main reason that some libraries conducted all or part of their interlibrary borrowing outside established state networking channels was their perception of network performance. And for libraries that had been actively involved in interlibrary lending for a long time and had established good working relationships with other libraries this had established good working relationships with other libraries this perception unfortunately proved all too often to be accurate. The multitier structure of most state networks could not help but result in processing delays of requests not filled at a lower level of the hierarchy. This could be particularly frustration when a library was requesting an item known to be at one of the upper-level referral sites and extremely unlikely to be found at any of the intervening levels.

In such a case the originating library might reasonably feel that the request should be sent directly to the referral library that was known to own the item, but mechanisms were set up in many state networks to discourage this practice. Compensation for in state lending was sometimes paid to referral libraries only for loans processed through normal state network channels, and in many cases referral had a policy of not accepting direct requests from in-state libraries, at least of less than a certain size. These state networks created during the mid-and late 1960s had little on which to draw in the way of precedent.

The environment in which they were established was characterised by an increasing level of interlibary loan activity and an inadequate level out title-specific locational information in distributable from. These two conditions worked toward opposite ends and a highly structured communications network was seen as the most appropriate available mechanism for coping with the circumstances. Despite their sometimes cumbersome operation, state networks were not unresponsive to change, and the agencies responsible for their administration tried in many cases to streamline operation, broaden the scope of materials that could be borrowed, and incorporate new procedures or policies as need arose.

Perhaps most importantly, the networks provided a place to which many local libraries with little or no previous experience in interlibrary loan services could turn for assistance and a formal mechanism for those libraries to gain access to a far broader base of resources. Two years after the creation of NYSILL, S. Gilbert Prentiss, former New York assistant commissioner for libraries, said of the performance of NYSILL and similar networks: "The mere establishment of a system almost inevitably builds in a degree of inflexibility and cumbersomeness which in certain situations will be comparatively slow and costly even though it expedites services to most users most of the time.

Against a background of increasing level of interlibrary lending state interlibrary loan communication networks carried our important functions as agents for regulating the flow of requests within a state and for providing a structure for locating requested items. These functions were needed in part because of the lack of comprehensive tools providing known locations of books, serials, and other items that might be requested through interlibrary loan. The hierarchical structure off most networks provided a mechanism for locating items in at least some library whose holdings were not likely to be reported in such tools as the national Union catalogue thereby diverting some requests from those libraries at the top the hierarchy whose holding were available on a wider basis. It is uncertain more libraries would have entirely obviate the need for the formal network established in many states.

It is certain, however that the lack of such knowledge was a major obstacle in the interlibrary loan process. Although more prolific in number locational tools available even by the late 1960s had changed little in form or nature of complication from precedents established during the first half of the twentieth century. Although more and more libraries had been gaining experience in the automaton of internal operations through the use of computers, and unit-record equipment before that, there had as yet been little application of these tools in coordinated projects involving several different libraries. Union catalogue of a sort had been produced on unit-record equipment as early as the 1950s, but these were invariably such as a public library or regional system and it is branches.

In several instance, even through the 1960s, these catalogue contained only bibliographic information about items held within the system and information on specific locations for each term remained available only on a central union card catalogue shelflist. The major difficulty in automating interlibrary union projects was the few libraries had bibliographic records in machine-readable form and even those that did were almost certain to have recorded them in a format deferent from that used by any other library. As a result machine-readable records created by a library for its own internal operations could not easily be combined with those of another library to establish a common database for generating a union list or catalogue.

The automation of union project there fore, had to be well through out and coordinated in advance to ensure the greatest possible compatibility and consistency in format; it could not be carried out as an afterthought. A few interlibrary union project were initiate during the early 1960s using unit-record equipment. One focusing on monograph holdings, was a joint project involving the Niagra Falls and Lockport public libraries in New York in 1964. The scope of the catalogue was fairly limited and did not entail conversion of records for items existing in the collections before that start of the project.

Also in the mid-1960s the Rio Grande chapter of the Special Libraries Association (SLA) issued a union list of serials, printed on unit-record equipment to which about 30 special libraries contributed. In a similar venture, the Minnesota chapter of the SLA generated a 130-page union list of serials in 1963, containing about 3,500 entries. As with other library applications, unit-record equipment had limited applicability for the preparation of union lists and catalogue because of the speed at which it operated. Already by the mid-1960s, a few groups of libraries were beginning to engage computer for union listing projects.

One of the earliest was realized in the 1964 publication of the 13,000-entry North Carolina Union Checklist of Scientific Serials, containing titles and holding information for scientific serials in five university libraries in the North Carolina Research Triangle. The Kansas Union List of Serials, 1965, contained more than 17,000 titles and reported holdings information scientific serials in five university libraries in the North Carolina Research Triangle. The Medical Library Center in New York City began in 1964 to create a machine-readable database of holdings information for serials titles in a number of medical libraries in the city.

In 1967, the first edition of a printed list, Union catalogue of Medical Periodicals (UCMP) was issue, containing the more recent and active titles with holdings information for more than 50 medical libraries in the metropolitan area. These examples were representative of computer-based union projects during the

1960s. Most were targeted exclusively toward periodicals, or serials, and encompassed libraries limited either by type of geography. During the 1970s, and often backed by LSCA Title III funding, a number of states initiated computer-assisted union listing projects. In some cases, these statewide efforts were based on previous or existing projects of lesser scope.

In Kentucky for example a title checklist for collecting holdings information from a large number of libraries in the Louisville area and in southern Indiana. Data from the returned checklists were converted to machine-readable form and stored on a computer at the University of Louisville. The original intentions was to send checklists and forms for adding titles not yet on the lists to about 100 of the nearly 300 libraries in Kentucky. This number was eventually pared to 51, but it represented a range of academic public, and special libraries.

In 1972, the first edition of the computer generated 16,000 title Kentucky Union List of Serials, with about 37,000 holdings was printed and distributed. As early as 1968, funding had been appropriated in Indiana for preparing a broadly based Indiana Union List of Serial using computer facilities at Purdue University in West Lafayette. Considerable difficulty was encountered along the way but in January 1970 a preliminary computer-printout listing was issue, containing about 46,000 entries with nearly 50 libraries in the state contributing. The reception was not overwhelmingly enthusiastic; criticism was aimed primarily at the format and filing order in which entries were listed.

A second preliminary catalogue issued in early 1971 was not much more favorably received. After a great deal of reworking of data and soft ware programs, the Indiana Union List of Serial First Cumulative Edition was issued in late 1973, containing more than 90,000 entries will more than 150,000 holdings. Minnesota also mounted a concerted effort to complete a broadly based union list of serials. With responsibility for coordination awarded to MINITEX the Minnesota Union List of Serials (MULS) project was formally initiated in 1971.

A preliminary two-volume edition containing about 37,000 entries and 20,000 cross-references and added entries, was issued a year later. The first official edition was published in late 1974 in five volumes with more than 70,000 titles and contained holdings for 122 libraries. The machine-readable MULS database served as one of the three serials files, along with those of the Library of Congress and the National Library of Canada, used to create the initial CONSER (Conservation of Serials) file loaded into OCLC. The generation of a broadly based union list of serials from machine readable data represented a substantial undertaking in the early 1970 but it will placed in comparison with the effort and expense required to generate a union catalogue that would have included bibliographic and holdings information for monographs and other materials for the same number of libraries. In 1968, the Connecticut State Library announced its intention to coordinate such an effort to be backed by an inclusion of $500,000 of LSCA funding over a five-years period.

By 1971 however, the project had been abandoned and was replaced by a plan to microfilm various card catalogue and store copies at the state library as tools for assisting libraries in finding interlibrary loan locations. Compiling a machine-readable database representing monographic holding of a number libraries posed some serious problems that would have had to be solve pertaining to uniformity in bibliographic format, magnitude of converting manual records to machine-readable form, and the enormous amount of computer space required to store the records.

A few states, however, invested in a less ambitions form of union locational tool known as the numerical register. Rather than indicating full bibliographic information for each item held by reporting libraries the entire "catalogue" consisted of Library of Congress Card Numbers (LCCNs) and location symbols showing which libraries owned an item represented loan request was submitted by a library user, its LCCN could be determine through a number of sources, such as the National Union catalogue, in which LCCNs were provide as part of the bibliographic entry. These numerical union registers, such as the one published in Louisiana in 1971, did circumvent the problems of bibliographic uniform and did reduce conversion effort and the amount of computer storage was restricted to items catalogued by the Library of Congress, and a mechanism was require to assist smaller libraries that might not have adequate tools for determining the LCCN of an item requested by a patron.

Of all the efforts during the late 1960s and early 1970s to produce monographic upon catalogue with the assistance of a computer and a machine-readable database, the most ambitious was the University of California's Union catalogue Supplement. Published in 1972, it was a 48-volume 42,000 page book catalogue of all monographs catalogued by libraries on all nine university campuses between 1963 and 1967. The time effort ad expose involved in such a project were out of the question for most states and university systems to even think of attempting a similar feat. It was perhaps a fitting monument to the end of an era that the Supplement was one of the largest-and one of the last-computer produced monographic union catalogue to be published in book form. Large union catalogue of monographic holding would be published in the future, but their medium would be computer output microform rather than paper. They would also invariably be based in large part of the off line products of a type of union catalogue emerging in the early 1970s, the online union catalogue.

One of the earliest online union catalogue in the United States, and the one that established itself by the mid-1970s as the most broadly based in American libraries was OCLC. Its emergence in an online mode in 1971 as a share cataloguing system soon had important implication for interlibrary loan. Each time any members library catalogued an item using on OCLC bibliographic record, its cation symbol was affixed to that record. Although many libraries joining OCLC had been reporting holdings to the National Union catalogue the most majority had not, For most, OCLC was the first opportunity to have their monographic holdings known to other libraries.

A large number of libraries chose not simply to catalogue their current acquisitions on the system, but also entered holding for materials already in their collection, before joining OCLC. As the shared database grew through the 1970s and as more libraries joined the system. OCLC evolved into one of the most powerful interlibrary loan locational tools available. In some ways, the locational aspect to the OCLC system competed or conflicted with the role that had been carve out for state interlibrary loan communications networks. It represented an alternative for finding interlibrary loan locations. Point-to-point communication between laborers relying on OCLC for holding information even if locate in different states, often meant that request were handled more promptly than through state network channels.

In the most extreme cases OCLC was seen by certain agencies as an intruder impinging on an already established area of responsibility and as an alternative threatening to reintroduce the unchanneled, free flow of requests that had typified interlibrary loan before the creation of the state networks. In many other cases, however state networks were more open to OCLC and incorporated it into the network structure as an additional tool for locating items both within and outside the state. OCLC and the other bibliographic utilities that emerged later not only provided online union catalogue of locations, but they also enabled participating libraries to build their own machine-readable databases supplied in magnetic tapes.

The different between these institutions specific machine-readable databases and those created by libraries during the mid and late 1960s was in the degree of format compatibility from on library to the next. The credit for this compatibility rested in the widespread adoption of the MARC II format by libraries and bibliographic utilities as a standard for machine-readable representation of bibliographic data. This standardization allowed the information on one OCLC library's magnetic tape to be merged with another's with a minimum of efforts, or even the records of an OCLC library with the records of an RLIN or WLN library, and so forth.

There were minor problem in merging due of slightly differing interpretations of the MARC format utilities and even individual libraries, but in essence the records were compatible. During the mid-and late 1970s a number of states took advantage of this compatibility by merging records of libraries with MARC based machine-readable files into a single statewide database to be used in a union catalogue served as an indispensable tool for smaller libraries unable to afford membership in a bibliographic utility, and in some states it bridged the gap between libraries using different utilities or other systems that could not be cross-examined in an online mode. Alabama, Colorado, Florida, Kansas, Maryland, Mississippi, and Nebraska were among the states that developed stateside COM catalogue by the early 1980s, and many are still in existence and are updated periodically.

The 1970s was the most active decade in the history of American libraries in the distribution of holdings information on a nation wide scale, and this activity continued into the 1980s. By mid-1984, the OCLC system alone continued more than 150 million holdings for its nearly 11 million bibliographic records, and there were million of other holdings in the RLIN and WLN database. In addition to the birth and growth of online bibliographic utilities during the 1970s, the 700-plus volume National Union catalogue: Pre-1956 Imprints was completed. In the early 1980s, several union lists and union listing was capabilities were brought online. The California Union List of Periodicals become available through BRS; the OCTANET combined serials check-union listing and interlibrary loan system was activate; and OCLC union listing capability was implemented and by mid-1984 contained more than 50 local state and regional online union lists.

Activities on regional and nationwide systems were supplemented by efforts at the state and local levels. With the assistance of computers, a multiplicity of localized union lists and catalogue was being generated. Another important development was the evolution of clustering arrangements in automated circulation systems and online catalogue, where libraries share a central computer and are also all to search each other's online file.

Loan Requests

During the 1970s when the distribution of holding information was progressing at a rapid pace, there was little change in the techniques used to communicate interlibrarial loan requests and associate correspondence between libraries. More libraries installed TWX machines or similar equipment, but he changes were of magnitude rather than kind. During the litter half of the 1970s, thus began to change as some existing online systems introduced enhancements to link two previously distinct parts of the interlibrary loan process: identifying locations and communication requests. The earliest precedents were found in recommunication requests.

The earliest precedents were found in remote-access circulation systems, such as the system developed for use by the 21 libraries on the campus of the Ohio State University. In 1979, two existing systems were expanded to provide a link between location identification and transmission of request in networking environment. One system OCLC had existed in a networking environment for several years, but had not had an interlibrary loan communication capability. The other system the Library Computer System (LCS) in Illinois, had not previously existed in a networking environment beyond a single campus, but had already incorporated a link between identification of locations and transmission of requests to borrow.

Before 1979, the OCLC online catalogue was an important finding tool for interlibrary loan locations, but once locations were identified, the request itself had to be transmitted via other means mail, telephone, or TWX. After a three-month test among 20 libraries in early 1979, OCLC introduced alternative to meal telephone and TWX. With this new capability once a library retrieved an OCLC bibliographic record for an item it wished to borrow, it could select up to five potential lenders from among the OCLC membership.

By keying in a simple command, it could then request an online interlibrary loan work form for the item. Bibliographic information from the OCLC record was automatically transferred into the work form along with the identifier symbol of the libraries designated as potential lenders. The borrowing library then merely had to complete the work form by keying in its address and, if it wished other information such as a patron's name. When the online work form was completed a sample command sent it to the central computers in Ohio and immediately into an online message file of the first potential lending library specified by the borrower.

The potential lender received the online request and replied that it either could or could nor supply the item. If the lender responded that it could not the online request went into the message file of the second potential lender, and so on. When a lender responded that it could supply the item, the request was subsequently updated online by the borrower and lender at various points of the transaction to indicate its precise status at any given time; shipped, received, renewal requested, returned, and so on. Immediately upon its introduction the subsystem was well received by the OCLC membership.

In most instance, it enabled the library to combine bibliographic verification, identification of holdings, and transmission of a request into a single continues step requiring only one session at the same terminal.These features, combined with instantaneous transmission, the ability to designate a routing chain of up to five libraries, and the automatic referral from one potential lender to the next after a negative response, reduce the elapsed time between request initiation and document receipt for most libraries, and resulted in a higher overall fill rate than had previously been the case inmost libraries.

In all the Florida Library Information Network (FLIN) study the fill rate consistently exceeded 90 percent. Another consistency in these studies concerned the distributed of filled requests by position of potential lenders in the lender chain. Except in the first New York study, the first lender to which requests were sent filled between 52 and 57 percent of all requests that were eventually filled, and the second lender accounted for between 21 and 29 percent. This compared with an average of 15.9 days for requests sent through the mail during the preceding two years when OCLC was used only to identify locations.

In the first New York study comparative averages were 17.0 days for requests sent via OCLC and 27.4 days for those sent through the mail during the same period. Comparative averages for the second New York study were 16.3 days and 22.1 days respectively. Use of the OCLC communications capability remained largely outside formal state networking channels. Just as there had been rather reservation about how OCLC as a locational device would affect the established channels of communications within a state network, so was there a reluctance after 1979 by many state systems to last the communications capability of OCLC.

By the late 1970s many sate networks had finally incorporated OCLC as locational tool, playing terminals in some or all of their systems and resources mods. But when the communication networks to continue to use OCLC only as a locational tool, relying on previously established modes of communication such TWX for actual transmission of requests. There were some early exceptions to this general tendency, such as in Texas, where the state networking system adopted OCLC as its primary interlibrary loan communications system.

A second major online system to combine holding information with transmission of requests in 1979 was the Library Computer System (LCS) of the University of Illinois. LCS was not an interlibrary union catalogue like OCLC but rather a circulation system on which initially the file of one library, and later others, were available of remote searching and charging. The initial database on the system was comprised of records for most of the holdings of the University of Illinois Urbana Champaign libraries. In December 1979, hard-wired LCS terminals were placed in each of the 18 ILLINET regional systems headquarters.

Terminals were later placed in a number of other locations in Illinois and several libraries in the state loaded their own machine-readable databases into LCS and were using the system for circulation control. The database of the various libraries were maintained separately on LCS. Each library, however, was able to search the holding of the other libraries and could directly charge out items online from them. When a charge command was entered, a page slip was printed at the lending library and the materials was retrieved an sent to the borrowing library.

A real advantage of LCS was that each online bibliographic record displayed the circulation status of the item. With this information the borrowing library had a fairly good idea as to whether the item was indeed available for immediate loan by the library whose file it was searching. Other systems had integrated functions of location identification communication, and circulation charging before this time, but they were largely in libraries under a single administration or within a single jurisdiction. A unique aspect of the Illinois experience was that the LCS participants were administratively independent libraries located throughout the state.

System providing a communications as well as a locational capability helped to streamline interlibrary loans. In 1981, enhancement to the RLIN system enabled its members to more quickly and easily identify all recorded stations among RLIN libraries for any title in the database. In 1982, a sophisticate online interlibrary loan communication capability was introduced. Another development along different lines but also receiving considerable attention was the emergence of electronic mail services (EMS), such as TYMNET's One-Tyme II system. These have a wide spectrum of potential applications for libraries.

In terms of communication for interlibrary lending, they are less sophisticated than the interlibrary loan subsystems of utilities such as OCLC and RLIN Messages can be sent to one location at a time or to several simultaneously, but stepwise referral from the one lender to another is semiautomatic at best and bibliographic information has to be keyed in its entirety into a request rather than any portion being automatically transferred from an outside source bibliographic record.

Document Delivery Services

Electronic mail services shared circulation systems and the interlibrary loan subsystems of some of the utilities were all playing an important role in interlibrary loan by the early 1980s. But even with the improvements in locating materials and transmitting interlibrary loan requests a persistent problem remained the matte of actually delivering the documents themselves. Improvements on the time factor in documents delivery during the 1970s were the result of better organization and exploitation of traditional means of delivery rather than the results of technological innovation.

Both on an informal and formal basis, increasing numbers of libraries in local cooperatives and state networks began to exchange materials via courier delivery. Some of the information retrieval services and databased vendors also began to offer document delivery services. Although copies of documents were still sent through the mail by these services, a high priority was placed on rapid handling of orders and mailing via the fastest postal class a available. Although there had been some experimentation with telefacsimile transmission throughout the 1970s, it was isolated and fairly narrow in scope.

In 1979 however two important experimental project was the Federal Library Network Prototype Project (FLNPP) in which 12 federal and 3 nonfederal libraries in various parts of the United States participated. The other was the Telex Library Information Network (TALINET), with five sites in the Midwest and Rocky Mountain regions, which were experiencing rapid population and economic growth a result of an influx of energy-related industries. Several different type of equipment and capabilities were tested during these experiments. Some transactions involved interlibrary loan requests for copies of articles; others were requests for answers to reference questions, still others were requests for online literature searches.

In general, the TALINET and FLNPP project team felt the results were favorable. Since the FLNPP and TALINET project in 1979, there have indeed been important development in telefacsimile technology. Machines are now widely available that a transmit at a rate of one minute per page or faster, and the cost of equipment has decreased considerably. Although telefacsimile technology appear to be rendering more favorably for libraries and the service is becoming more cost effective there have been, and remain, some very fundamental questions about the suitability of telefacsimile for the transmission of documents in interlibrary lending. One question concern the level of need. It seems an almost irreproachable tent in the library profession that it is desirable to deliver information to the library user by the fastest means possible.

Yet in the 1967 CLR-funded telefacsimile experiment, library users frequently did not pick up materials for two or more days after they had arrived. In the New York FACTS project, more than 70 percent of the respondents to a questionnaire indicated that the material would have been just as useful to them had it arrived two or three days later. In the TALINET project, only about 28 percent of the requests were "time critical" answers to the majority of requests" could have been handled by conventional means (mail) and still have been useful". A significant factor in these user responses of course, may have had to do with learned levels of expectation. It has always been true that interlibrary loan does not provide same-day delivery, or sometimes even same-week or same months delivery, and library users have grown to accept or work around this fact.

If this is needed a significant factor, then one cannot discount the importance of or need for more rapid delivery of documents for at least some types of requests. Beyond the question of user expectations, however, is an equally important consideration: the processing priorities assigned by the libraries involved. The actual time spent in transmitting a copy of a document can be greatly reduced through the use of telefacsimile equipment, but the mere substitution of one mode of delivery technology for another will not in itself insure what some individuals would perceive as rapid document delivery.

7

Bibliographic Display

In a few OPAC, an added preliminary procedure is the selection of one of two or more optional major command modes-for example, beginner or advanced. Another preliminary action on some OPACs is file selection, which usually occurs in two circumstances: in multipurpose systems in which the public catalogue is only one of several files accessible to the user, or in multi-institutional POACs in which several libraries catalog are independently accessible. In either case, the user may have to enter a command to active the appropriate file or to switch from one file to another before a search statement can be entered.

Relate to but less comprehensive than specifying entirely different command modes, another preliminary begin-session procedure found is some OPACs enables the user to specify certain limits related to the language or form in which dialogue will occur, most often having to do with system response. These options dictate such characteristic as the length of system message or the format in which single bibliographic record are display. For example, the system might allow the user to specify in advance whether certain types off system messages displayed will be brief in content or more fully explanatory. Additional begin session procedures are usually options, not requirements. For each, there is almost always a "default value", a single option under which the system operates unless explicitly told by the user to do otherwise.

Public-access terminals in multipurpose systems, for example are normally present for searching the public-catalog file. As long as the user does indeed wish to search that file, no explicit action is required to select it. In multi-institutional OPACs maintaining a separate file for each library, the terminals in each library are usually set for searching that particular library's catalogue. Only if the use wishes to search a catalog of the other libraries is it necessary to use file-section procedures. OPACs that provide different message lengths or bibliographic record formats usually operate the same way. One message length or record formats usually serves as the default value, and a user must make a preliminary selection only if an option other than the default value is desired.

In general, OPACs keep preliminary steps to a minimum, and additional procedure cater to the more experienced user or to those wishing to exercise an option that is different from the one judge to be of most use to the majority. The relationship between search statements and retrieval characteristics is perhaps the single most crucial link in a dialogue between a user and an OPAC. Although this relationship may at first seem self-evident and straightforward, that is not always the case, Many characteristics differ in application from one OPAC to another, and not necessarily in the most obvious ways.

After carrying out the necessary begin-section procedures, the next stage entails the entry of a search statement. Before nothing the variations for construction of search statements, it is necessary to see how search types relate to two other concepts: access points and indexed fields. The relationships among these three are crucial in considering the formulation of search statements and in determining how those statements interact with the retrieval characteristics of a system. The three terms-access points, indexed fields, and search types can be confusing. They are sometimes used almost interchangeably, and one source of confusion is that, in some OPACs the concepts they represent do in fact rather closely coincide.

In others there are very important distinctions among them. Shows three examples of relationships among access points as defined in the traditional sense of the term, indexed melds as applied within context of machine-readable bibliographic records, and search types as employed in some online catalog. These examples do not cover all the ways that the three concepts relate to one another in all OPACs, nor do they address the question of numeric search types (LCCN, ISBN, and so on).

However, each example does represent a relationship found in at least on OPAC now in use. In the first example in there is essentially a one-to-one correspondence among access points, indexed fields, and search types in an OPAC. An author search allows access by name, retrieving records based on a match between the author-search terms entered by the user and terms appearing in designated author fields on machine-read-able records. The same correlation applies to the and subject searches. In some OPACs, the search statement must include the full author's name, exact title, or a complete form of subject heading as it appears in machine-readable records.

Other OPACs are more flexible, allowing for retrieval on keyword and truncated terms searching. But even with greater flexibility, there is a strict correspondence among access points, indexed fields, and search types. This is not always simply a change in nomenclature, for there might be a somewhat different set of indexed fields that serve as the basis for retrieval. For instance, personal and corporate name subject headings might be added to the list of author fields indexed for searching by name. Most visibly it eliminates title and subject searches as distinctive types, encompassing them both.

In addition, some OPACs that use this more comprehensive term-search approach index fields other that title and subject for retrieval. Usually this means the addition of notes fields, but some OPACs include others as well. The system will retrieve all bibliographic records in which search terms entered by a user appear in the title of a work, in an assigned subject heading, or in any other field that might be indexed for term searching in that particular OPAC.

The third example in points up the most radical approach to the relationship among access points, indexed fields, and search types. It revolves around the designation in of only one search type, thereby virtually eliminating traditional distinctions among author, title, and subject searching, and it does not even recognize the distinction between names and other terms. These three examples represent different approaches, but a number of line catalog incorporate more than one approach. Several OPACs that provide distinct author, title, and subject search types, for examples, also have a more comprehensive term search that acts of various combinations of these fields as well as perhaps on other, indexed fields.

An interesting and somewhat unusual approach to this whole matter is used in the online catalog of the Beth Israel Hospital Library in Philadelphia. Search terms can be entered on this system without search type designators, and the system will retrieve records in which the terms appear in any indexed field on the record. If however, the search retrieves a large number of records, the user can then qualify whether the entered terms should apply only to author, title, or subject fields.

There are a number of different approaches used in OPACs for constructing a search statement. After the required begin-session procedures have been carried out, the system is prepared to accept and process a search statement, which generally consists of a search command name, a search type qualifier, and one or more search terms. There are significant differences among OPACs in the specific procedures required to enter a search statement and these differences can be viewed in terms of three important implications:

— The extent to which all three components of a search statement-search command name, search type qualifier, and search terms must be explicitly keyed by the user.

— The number of steps required to enter the search statement.

— The flexibility with which a new search statement may be entered at various points during a dialogue.

One approach to constructing a search statement requires a user to explicitly enter all three components of the statement: search command name, search type qualifier, and one or more search terms. For instance, with the MELVYL system at the University of California.

In most OPACs using these methods, after the required return commands have been executed, the system is then prepared to accept a search statement. Thus once the point is reaches where a search statement can be entered, it is nor necessary to explicitly key in a search command name, as in the first two approaches. But remember that the user must execute one or more preliminary steps before reaching that point.

The third a approach to constructing a search statement, illustrated in relies on a fill-in-the-blank technique. From the introductory screen display, a user can enter a search statement in a single step. OPACs using this approach generally do not even require the user to explicitly state the search type. It is designate simply on the basis of where in a range of options, the search terms are entered, In the example in, after the user carried out any necessary procedures to retrieve the display, he or she could enter a search statement for the author, Murl Smith, simply by keying the searching terms following the system-supplied "Author": search type designator.

With the fourth basic approach, the user must carry out two separate steps in constructing a search statement once preliminary procedures are completed. In the first, step, the user selects the search type from available options. In some OPACs a mnemonic value is entered; others use a numerical value corresponding to each search type. The system responds to this selection by displaying a new screen, and the search terms are entered on that screen as the second step in the process.

All four approaches described here require the user to explicitly key in at least the search terms for which on OPAC will retrieve records. But some OPACs do not require explicit keying of even these elements. In the touch-sensitive version of CLSI's Public Access Catalog for example, the user virtually never keys and commands or terms onto a keyboard. Instead, the most user continually selects alternative commands or terms form a list or options presented on a screen simply by touching the screen at the place where the desired option appears. Although on the fourth approach described above.

A user begins a search from designated initial divides the entire alphabet into several groupings. Once the grouping that contains the desire term is selected, a new display breaks that alphabetical grouping into subgroupings. This narrowing process continues until the desired search term appears as a category in itself. As is readily apparent this single "step" for selecting a search term can easily consist of a rather large number of narrowing sequences and can be quite time consuming. In the construction of a search statement, all but a very few OPACs follow one of the four basic approaches described above. The "best" one for any particular library will depend on number of variables, for the entry of a search statement is just one aspect of dialogue between user and OPAC.

Many OPACs follow different distinct search type-such as a another, title, subject or name term, and so on-to be combined by the user in a single of the database. It was note that a Boolean searching capability enhanced access to bibliographic records online versions of the library catalogue. The ability to combine search types is a form of this capability. One OPAC that has used this approach is Brodart's online catalog illustrated in the same screen display can be used to enter either a single type search statement or one in which two or more search types are combine. In combining search types, the user merely enters each term in the space adjacent to the system supplied search type labels.

A third general technique requires the separate entry in one or more step, of each search type qualifier/search term part being combined, in obvious contrast to the first two approaches, in which pairs are enters in a single step. The user must execute two separate steps for each search type qualifier/search term pair: the first to indicate the search type, and the second to enter the actual terms themselves. This procedure is repeated for each search type qualifier/search term pair being combined, and thus will require several steps simply to combine two search types and the desired search terms. A few do provide additional combined search types. MSUS/PALS, which offers discrete author, title keyword, and subject search types as well as a combined author/title search, also provides a term-search type which essentially combines title and subject searches.

Several OPACs offer a term search capability in lieu of distinct title and subject search types but only a handful such as MSUS/PALS offers it in addition to the other two. Through there are variations on each, the four techniques described here constitute the major approaches used by OPACs allow a combination of distinct search types in a single search of the catalog. Each technique has advantages and disadvantages depending on user needs relate to simplicity, clarity, and expediency. All are fairly straightforward in execution, and only the third approach requires a substantially greater number of steps than those needed for entering single-type search statements.

Although enhancement to all traditional channels of access in library catalog is an important attribute of OPACs the broad concept of subject access merits particular attention. Prior to the 1980s it was generally held that known-item author and title searches were far more prevalent in library catalogues than were subject searches, a contention supported by numerous studies of card catalog use.

But in the early 1980s studies of use patterns in a number of OPACs revealed a different picture. Subject searches, at least in these online versions of the catalogue were the predominant type of search conducted by users. Subject access in card and offline, computer-generated catalogues is usually hampered by:

— rigidity and complexity of headings
— need for the user to know complete and exact form of headings
— minimal number of headings assigned to items catalogued into the collection.

Many features have been introduced into online catalog to broaden subject access and expand the flexibility with which records may be searched by subject. Different OPACs employ these features to varying extents. The general usefulness of each of the subject-access features outlined below varies, but the quantity in which they are available in any given OPAC is likely to have at least some bearing on the quality of subject access provided by that OPAC.

— User must know and key in exact, full form or assigned subject heading.

— Assigned subject headings are usually drawn, with local variations, from standard lists such as Library of Congress Subject Headings, Sears, and so on.

— Has no greater flexibility for subject searching that ward or COM catalogues.

Keyword searching truncation on assigned subject headings

— Both features, although not interdependent, introduce greater flexibility and can result in broader access than the character-by character exact-march approach described above.

— The following example for the single subject heading "Learning psychology of-Congress "compare how records must be retrieved using character-by-character exact-match searching with how the same records may be retrieved using keyword searching, subject heading phrase truncation, and combination of keyword and term truncation.

Character-by-character exact match:
learning, psychology of-congress

Keyword
learning
psychology
psychology learning
learning congresses
congresses psychology
etc.

Phrase truncation:
learning
learning, psychology
learning, psychology of

Keyword searching combined with term truncation:

psychol
learn
psychol learning
congress psycholog
psych learn con
etc.

Increased number of bibliographic elements indexed for subject retrieval

— Subject access expanded through concept of "term" searches.

— Records retrieved when search terms appear not only with assigned subject headings, but also within other designated elements-titles, contents notes, and so forth.

— Some OPACs offer both subject search and term-search types; former usually applies to assigned subject heading specifically; the latter is more comprehensive and includes other bibliographic elements, such as title fields as well.

Keyword Searching on Title Fields

— More restrictive than multiple-field term searching described above.

— This "backdoor" approach to subject searching draws on the principle that vocabulary in title of a work, especially nonfiction, normally is descriptive of work's basic subject content.

Shelflist Browsing

— Openly available browsing feature allows user to enter full or truncated call number and view records in shelflist order (in card or COM form shelflist files are maintained almost exclusively for internal bibliographic control housed apart from Public catalog, and seldom readily available to users).

— Allows review of works in shelf order (including titles in circulation or otherwise not on shelf) without physically going to stacks.

Online lists of subject headings

— In some OPACs the first of retrieval in response to subject search.

— Contains subject heading entered and subdivisions and or alphabetically proximate headings.

— User selects heading initially entered or otherwise listed system display bibliographic record assigned that heading.

— In some OPACs, retrieval of a subject heading list is a distinct to operating from retrieval of bibliographic records assigned a particular heading; one search type for entering heading to retrieve list of headings, another for entering it as search term to retrieve bibliographic records containing that heading.

— Currently primary emphasis for OPACs offering this feature is simply on providing alphabetical lists of appeared. headings and subdivision a few OPACs are attempting to incorporate more information about nonalphabetical relationships between terms, such as through see-also cross-references and explanatory notes.

Display of only subject headings from bibliographic records

— In the University of California's MELVYL system, a user may specify that only subject heading fields be displayed for a set of records retrieved in response to search.

— All bibliographic data other than the subject headings assigned to each record are suppressed in the display.

— This feature may be useful in guiding a user to subject headings closely relate to a topcoat on which some records have bee retrieved.

— In Biblio-Technique's BLIS System, this approach also allows for the extraction of all access points, including subject headings, for use in additional searches.

Because OPACs are such a comparatively recent phenomenon, the actual worth of many of these features for enhancing subject searching is still uncertain. Although equivalents of some of these features have long been offered in other forms of the catalog, they have seldom been so readily available to catalog users.

Libraries have long maintained card files in call-number order; many have displayed authorize lists, such as Library of Congress Subject Headings, near the catalog to aid in determining appropriate subject headings for a search; and many libraries include all subject tracings on all catalog entries for a particular item, giving clues to other headings where similar works might be found. But such features of card and COM catalogue have seldom been very actively promoted as aids to the library user. In the online catalog, such features are all available in one place, through a single terminal, and their presence is more apparent. Character-by-character and keyword searching were compare earlier in relation to subject access in online catalogues. But the contrasts between them are also applicable to access by little and corporate name and to a somewhat lesser extent, by personal name.

With a character-by-character approach, a search by subject, title, corporate name or personal name, or any other parameter, will retrieve only records in which the text in the appropriate indexed fields matches exactly the string of terms entered in a search statement. Keyword searching does nor require such a level of precision. Records will be retrieved as long as the search term or term appear anywhere within the appropriate indexed fields, regardless of position and regardless of what additional words might also appear in those fields. For any type of search, each approach has certain advantage and disadvantages.

The character-by-character method has some of the disadvantages. The character-by-character method has some of the catalog. The user must have complete information about a title or a form of name or subject heading as represented in the catalog. Keyword searching is not rigid, in that it enables a user to find records even if only a word or a few words in a title, name or assigned subject heading are known. However, keyword searching is not without limitations. Its flexibility is advantageous under many circumstances, but in some situations its impreciseness can a problem. An example of such a situation was described by Stephen Salmon as it applied to the prototype design of the University of California's MELVYL system.

Although magnifications were later introduced in the original version of MELVYL all alphabetical searches were essentially keyboard; there were no provision for carrying out character-by-character, exact-match searches. As a result a search of the database under "war and peace" in an effort to find records, for Tolstoy's 1865 work reportedly retrieved nearly 500 records, including not only those for War ad Peace, but also for all others in which the search terms appeared anywhere in the title.

The problem of imprecision can be especially profound when using broad term-search type that retrievers records based on the appearance of search terms in any of a variety of indexed fields. Although a number of OPACs incorporate both the character-by-character and keyword approaches, quite a few provide only, one and there is till an inclination in some cases to remote one or the other as being more effective or otherwise "better", In actually the greatest liability of either approach is when it is provided exclusively of the other. Nether method is always the most suitable, for their strength and weaknesses vary according to circumstances.

For broadening the concept of subject access and for allowing users to find records when they possess only partial bibliographic information, a keyword search capability is indispensable. But especially for known-item searches in which a user does have complete bibliographic knowledge-an exact title or author's name as it appears in the catalog-a character-by-character search capability can be clearly superior. A keyword only search when full bibliographic data are known probably means that an inordinate amount of time and effort will be spent by the user in sifting through a long list of records that he or she known to be irrelevant to the purpose of that search. An OPAC should provide both character by-character and keyword search capabilities, and a number already do.

With both approaches the user is given maximum flexibility for selecting the search strategy that will be most suitable and most economical for his or her specific circumstances. Like keyword searching, truncation opens the possibility of broader retrieval than does character-by-character exact-match searching. The basic questions concerning truncation in an OPAC are whether it is available at all and if so, whether it applies to terms phrases or both, and whether it must be explicitly stated by a used or is implicitly assumed by the system.

The clearest example of truncation is in the use of derived search keys, whereby a user is required to truncate each search term according to a prescribed formula and is restricts to a maximum number of such terms that can be entered. Derived search keys, not common in OPACs have long been used for some types of searches in Ohio State University's LCS. A user conducting a title search on this system has had to enter the first five characters of the second word. Fewer characters could be entered if either of the first two words in a title contains fewer than the prescribed maximum, or if the title consisted of a single word, but a greater number of characters from the first or second words could not be entered, nor could characters from subsequent words in the title. So a title search for Quantitative

Techniques for Business Decisions would have to be entered as. If the desired item is in the database, it will be retrieved by using the above search key, but so will records for nay other titles that have "quan" as the first four characters of the first significant word and "techn" as the first five characters of the second. LCS also used derived search keys for author searches and for combined author/title searches. The derived search key example in LCS is very explicit in terms of user action: It is the user who must perform the truncation. Some OPACs however, have internal truncation; it the system and not the user, that formulates the derived search key.

A user may enter more complete search terms—for example, the complete title Quantitative Techniques for Business Decisions—but before processing the search, the system will internally truncate the phrase and each term according to a formula, that is not apparent to the user It is this truncated search statement, rather that the user's more complete statement, that serves as the basis for retrieval. The desired record will be retrieved in the database but so will others for which the internally truncated statement. A certain amount of imprecision is inherent in derived search keys. The search key algorithms can be designed to maximise precision, but there are bound to be some searches that retrieve a large number of records that are not relevant to the user's needs.

Perhaps even more of a problem with internally derived search keys, where the user cannot see the truncation that occurs, is that he or she may be confused about why the system is responding the way it is. Derive search keys, which demonstrate both term truncation (searching on less that complete words) and phrase truncation (searching on fewer words than might be included in complete bibliographic data), are extreme examples of truncation in OPACs.

In system that do not use such keys, term and phrase truncation are often approaches somewhat differently. Term truncation, when available is usually associated with keyboard searching and almost invariably required explicit action on the part of the user, generally in the form of keying a designated nonalphabetical character-a number sign (#), asterisk (*), or hyphen (-), for example as the last element of a search term.

Although explicit entry of a truncation designator by the user is the general rule for term truncation, a few OPACs do implicit treat search terms as indicated forms. In these systems entry of the term electrical would always retrieve records in which derivative word such as electrical and electricity appear in the title. There is no mechanism in this approach for the user to specify whether a search term is being entered in truncated or complete form; it always treated as a truncated term. Term truncation is most commonly associated with keyword sketching but phrase truncation is sometimes linked with character-by-character searching.

Some OPACs do not use phrase truncation in any form. Record are retrieved only from an exact match between term as entered in a search statement and terms appearing in the designated indexed fields of records. A title search consisting of the two terms social statistics will retrieve only records in which those terms make up the full and exact title. Records will not be retrieve for such a title as social in the eighties. A character-by-character subject heading search under the phrase clocks and watches will retrieve only records in which those three words the exact and full form of a subject heading. It will not act on subject heading in which clocks and matches is followed by subdivisions or other terms.

In OPACs that to incorporate phrase truncation, the same two basic approaches are available as for term truncation: explicit, in which the user specify truncation, and implicit, in which the system automatically assumes it. With an explicit approach, the user normally designates truncation by entering as the last element in a phrase a designated nonalphabetical character, similar to those often used for term truncation. If the special designator is not included, the system will treat the string of search terms as a full expression and will retrieve only records of which it finds an exact match in the appropriate indexed fields. With either explicit or implicit phrase truncation, the results is to expand the scope of retrieval.

For a title search containing the phrase social statistics phrase truncation will retrieve records not only of those works whose full title is Social Statistics, but also for such tilts Social Statics in the Eighties. A truncated subject heading phrase search under the phrase clocks and watches will retrieve not only records in which that is the phrase cocks and watches will retrieve not only records in which that is the complete heading but also records for such headings as Clocks and watches-adjusting; Clocks and Watches, electric; and Clocks and Watch in art or any other headings in which clocks and watches are the first three terms.

As with term truncation, and explicit approach to phrase truncation is more flexible for the user because he or she was the option of entering a complete expressing or as a truncated phrase. The implicit approach leaves the user no optician; The sequence of terms will always be treats as a truncated phrase. As described earlier, this can lead to frustration for a user who knows the precise title desired or the full form of an author's name or subject heading, especially if only one or two words are involved. The desired records will be retrieved, but probably so will a long list of others that have little or no relevance to the user.

Truncation can be valuable enhancement to searching in an online catalog. Like keyword searching, however, it is suitable in certain situation, and can be a deterrent in others. And it should be provided not instead of character-by-character exact match searching but as an additional option. Its use should remain at the discretion of the user, not as a mandate of the system. In any OPAC, some proportion of searches will in no records being retrieved. This proportion can be significant.

For example, in studies in the early 1980s Northwestern University reported no retrievals for 39 percent of all subject searches in its LUIS system; the University of California note 35 percent; and Syracuse University gave 53 percent for no retrievals in the online catalog.

Many factors contribute to these high number but obviously a main one must be the discrepancies between user and catalog terminology to represent the same concepts. Character-by-character exact-match searches on names and assign subject headings are especially prone to such discrepancies, where the user enter terms in the exact form as the library has chosen to represent them in the catalog. Some OPACs have not yet addressed this problem through any online mechanisms. In these OPACs discrepancies between user and catalog vocabulary simply result in no retrieves, and it is up to the user to guess what other terms might be appropriate or to seek official guidance in determining the proper terminology.

The first method focuses on online display of cross-references. When the user enters a search term contained as a cross-reference in the authority file, the system refers the user to the library-established from form of heading used for that name or concept in the catalog. An example, cited by Charles Hildreth, is as follows:

— The name you entered, Holt, Victoria, is entered in this catalogue under Plaidy Jean.

— Do you wish to continue this search? (If "Yes" press the RETURN key).

The user must then either key in the proper from or as in the example above, simply enter a designated command to search under the correct from. The second method is to provide a strictly internal link from an unauthorized heading to an authorized form, displaying the correct bibliographic records with no further action by the user. This eliminates some of the user's work but can be confusing it a use enters one name or subject heading he or she may conclude that the system misunderstood the search terms.

The way in which an OPAC treats alternative terminology is especially important because it deals a very complex situation. The quantity of alternative terminology that can be used to express a single concept, especially for a subject, can be significant. One of the more promising techniques developed to help reduce this problem is to program an OPAC to record the terminology used in searches that retrieve no records, and then to analyze that terminology in order to add new cross-reference to the authority file. When that occurs, the next major concerns in the dialogue between user and OPAC are the manipulation of result and the presentation of bibliographic information.

Manipulation of results includes procedures to broaden the set of records retrieved, narrow the set, or display one or more specific records within the set. Rather than discussing in detail the various techniques for manipulating results, the primary focus in the remainder of this chapter is on the presentation of individual bibliographic records in terms of the information they contain. Just as there are difference among OPACs in the procedures required for searching so there is variety in the presentation bibliographic information retrieved in response to searches. This section focuses on two levels of presentation:

— first level for those instances in which a search has retrieved multiple records, and
— second level focusing on the display of single bibliographic records.

A search in an OPAC has one of three results: No records are retrieved, only one record is retrieved or more than one record is retrieved. If no records match the search statement, most OPACs display a message to that effect, and the user may enter a new search statement. If only one record is retrieved, the majority of OPACs will display a short medium, or long version of the record. More than one record is almost certain to be retrieved in searches on assigned subject heading, in keyword searches of various types, often in author searches, and sometime evens in title searches.

When multiple records are retrieved, the first level of presentation in most OPACs consists of a list containing one or up to a few lines of brief bibliographic data about each record matching the search statement. If the number of records is relatively small the list will fit on a single screen. For a large number several screens may be required in which case the user page or scroll to subsequent screen until the desired record or records are identified. Almost all OPACs use the brief-listing approaching to multiple retrievals, but there are alternative.

In at least two system, each will always result in the display of a single bibliographic record, even when more than one matches the search statement. These user page through subsequent records one at time. This is essentially the online equivalent of thumbing through a set of catalog cards in a drawer, one card at a time. Several OPACs instead of displaying the brief list require an intermediate step between the search statement and the display of any item specific bibliographic information.

The two most common intermediate steps are a display indicating simply the number of record retrieved in author names, terms or subject heading that match or nearly match the search terms entered. Whether in direct response to a search statement or as a result of an intermediate display, the first level retrieval of bibliographic records gives the user concrete search result in the form of selects bibliographic elements from each matching record. At this first level, there are distinct differences among OPACs in the amount of data supplied for each item when multiple records are retrieved. The basic different stems from the purpose for which this level of retrieval is intended. These purpose, and the degree of success in attaining them, can be assessed in terms of whether the information included about each item in a list is sufficient:

— to distinguish it from other items in the list

— for the user to positively identify it as the one that is though.

— to ascertain the location at which s should be found in the library

— to determine its availability status.

The display at the first level of retrieval for multiple matches in many OPACs is intended to serve only the fist purpose or perhaps the first two. To determine call number and other vocational data-and some-time even to make a positive identification of an item-the user must select a record from among the list and enter a command to view a longer version of it.

Other OPACs try to circumvent that need by displaying, for each item, more data at the first level, including more complete bibliographic description, locational information, and some-time even availability status. In some OPACs the data elements displayed for each items the first level of retrieval are fairly extensive including for example, Library of Congress Card Number author's full name complete title, full imprint statement, physical description, and call number. This amount of data should always be sufficient for a user to clearly distinguish among records included in the list, to positively identify the items desired, and to determine the shelf local of any item listed.

In order to include this much data for an item, its brief-listed record will have to be allotted between two and four lines of information on the screen display. But a far more predominant approach in OPACs is to restrict the description of each item in a multiple-record display to singe line. The "one-record one-line" approach does allow a greater number of brief records to be displayed on a single screen, and with enough information in most cases to distinguish items and to positively identify desired items. There are however glaring exception. Limiting the item description to single line means restricting the number of data elements that can be included, and it also restricting the number of character allotted to each element; obviously, in some case the data will not be sufficient for identification.

In searches that involve lengthy names of prolific authors or long and general titles such severe truncation can result in a listing in which items cannot clearly be distinguished from the number of data elements can have similar results. Some OPACs for example, did not include such elements as an edition statement or date of publication in single-line list at the first of retrieval. For a user seeking a record for a particular edition of a work that has gone through several editions, this limited first-level display will be inadequate to positively identify a desired item or even to distinguish item in the list from one another. In that and similar cases, the user only approach to select and view records in the list in their more complete version until the correct one is found. It some OPACs, user can through a single command, request the system to scroll longer version of several different records, but in most systems, they must select and display the longer version one record at a time.

Although the single-line approach does have some advantages, it can also be a real inconvenience. The data elements included at the first level of retrieval for multiple matches vary according to spare restrictions and more broadly, according to the purpose intended for this level of retrieval. The two elements most consistently displayed in nearly all OPACs are title and author. Invariably the title or some part of it still be include for each item in the list. If appropriate, an author's name is either included in each record or, specifically in response to author searches in some OPACs listed once at the top of list of items by that by that author.

The next most common elements are data of publication and call number. Other elements are far less frequent and are generally included only in those OPACs that devote two or more lines of description to each item. In some OPACs, especially those using the one-record one line approach, the data elements included depend on the type of search. The CTI online catalog for example, includes call numbers on brief records retrieved in response to author or title searches, but not subject searches.

The Ohio State University's LCS treats call numbers in exactly the opposite way: They are included on each record retrieved in response to a subject search, but excluded on each record retrieved in response to a subject search, but excluded in response to author, title or combined author/title searches. Others OPACs include the same data elements in all lists but arrange them differently according to the type of search conducted. In one system, for example, brief records retrieved from a title search call number. For subject searches, the order of the first two data elements is reversal: the data of publication appears first and the title second.

But in most OPACs, data elements and their arrangement do not vary from one type of search to another. Concerning the arrangement of brief records in a list with respect to one another, in most OPACs, the items appear in alphabetical or numerical order according to the first element displayed-author, title subject, or call number, and so on. Ideally, the first order of arrangement should defined by the search type alphabetically by author for author searches, by title searches and so on. As emphasized in the preceding section few OPACs include data elements beyond author, title, publication data, and call number at the first level of retrieval for multiple matches. This means that a rather long list of elements such as edition statement important, physical description, series statements, notes, subject headings, and added entries is left out.

Because these elements can be crucial for distinguishing between bibliographically similar works, for making positive item identification, or for helping to evaluate an item's relevance for a particular need, there are many instance in which users may want and need to see this more detailed information. All OPACs offer access to more detailed individual records.

In fact, most OPACs have at least two different versions or formats available for each item in the catalog, ranging from short record with a few elements of description to a fully ragged MARC record containing even more data than are traditionally associated with full catalogue entries. Most OPACs provide access to these individual records in one of two ways-automatic system response and user-initiated action. If a search statement is so specific as to retrieve a single record, most OPACs will bypass the one or two-line version and immediately display a more detailed form. Or from the brief- listing display when a search statement has retrieved multiple matches, the user may imitate the action by selecting an item in the list and requesting a more detailed record of it.

In OPACs that provide multiple version of a record for each item in the database, one format is usually established as the default value, and individual records for all items will display in that format unless the user explicitly requests a shorter or longer version. Of all format options available in all existing OPACs probably no two are exactly alike. There are so many data elements and so many ways to present the material that the precise form is highly specific to an individual system. Even so, some of the precise form is highly specific to an individual system.

Even so, some of the different are relatively minor, and at least at a general level of comparison there are many similarities. Two of the systems (TLS and LCS) have more abbreviated individual bibliographic record display formats available, and several allow an option whereby a user may explicitly request an even more detailed, numerically tagged MARC version of a record.

— data elements included

— whether data elements are labeled; if so, the terminology used

— sequence in which data elements are arranged

— spacing between data elements

It is evident in that, although the display are noticeably different in appearance, several contain a core of descriptive bibliographic data that very much mirror the types of information typically included on full-entry catalog cards. With some exceptions, OPACs generally have at least one display format-in addition to or other than a numerously tagged MARC format-that includes nearly all of the following type of descriptive data:

— Author

Title, including:

— Title proper

— Subtitle

— Remainder of title page transcription

— Edition statement

Imprint, including:

— Place of publication

— Publisher

— Data of publication

Physical description, including:

— Pagination

— Illustration statement

— Physical dimension

— Series statement

— Notes

— Subject headings

— Added entries

These elements are essentially the 100 through 800 fields of the MARC format.

TECHNICAL ISSUES IN DIGITAL LIBRARIES

The Internet would not be very useful if communication were limited to textual exchanges between speakers of English located in the United States. Rather, its value lies in its ability to enable people from multiple nations, speaking multiple languages, to employ multiple media in interacting with each other. While computer networks broke through national boundaries long ago, they remain much more effective for textual communication than for exchanges of sound, images, or mixed media — and more effective for communication in English than for exchanges in most other languages, much less interactions involving multiple languages.

Supporting searching and display in multiple languages is an increasingly important issue for all digital libraries accessible on the Internet. Even if a digital library contains materials in only one language, the content needs to be searchable and displayable on computers in countries speaking other languages. We need to exchange data between digital libraries, whether in a single language or in multiple languages. Data exchanges may be large batch updates or interactive hyperlinks. In any of these cases, character sets must be represented in a consistent manner if exchanges are to succeed. Issues of interoperability, portability, and data exchange related to multi-lingual character sets have received surprisingly little attention in the digital library community or in discussions of standards for information infrastructure, except in Europe.

The landmark collection of papers on Standards Policy for Information Infrastructure, for example, contains no discussion of multi-lingual issues except for a passing reference to the Unicode standard. The goal of this short essay is to draw attention to the multi-lingual issues involved in designing digital libraries accessible on the Internet. Many of the multi-lingual design issues parallel those of multi-media digital libraries, a topic more familiar to most readers of D-Lib Magazine. This essay draws examples from multi-media DLs to illustrate some of the urgent design challenges in creating a globally distributed network serving people who speak many languages other than English.

The technical issues involve the choice of character sets to represent languages, similar to the choices made in representing images or sound. However, the scale of the language problem is far greater. Standards for multi-media representation are being adopted fairly rapidly, in parallel with the availability of multi-media content in electronic form. By contrast, we have hundreds (and sometimes thousands) of years worth of textual materials in hundreds of languages, created long before data encoding standards existed. Textual content from past and present is being encoded in language and application-specific representations that are difficult to exchange without losing data — if they exchange at all. We illustrate the multi-language DL challenge with examples drawn from the research library community, which typically handles collections of materials in 400 or so languages. These are problems faced not only by developers of digital libraries, but by those who develop and manage any communication technology that crosses national or linguistic boundaries.

Medium, Culture, and Language

Speaking is different from writing, and still images are different from moving images; verbal and graphical communication are yet more different from each other. Speaking in one's native language to people who understand that language is different than speaking through a translator. Language translations, whether oral or written, manual or automatic, cannot be true equivalents due to subtle differences between languages and the cultures in which they originate.

The content and effect of messages are inseparable from the form of communication and the language in which communicated. Herein lies the paradox of information retrieval: the need to describe the information that one does not have. We have spent decades designing mechanisms to match up the expressions of searchers with those of the creators of textual documents (centuries, if manual retrieval systems are considered). This is an inherently insolvable problem due to the richness of human communication. People express themselves in distinctive ways, and their terms often do not match those of the creators and indexers of the information sought, whether human or machine. Conversely, the same terms may have multiple meanings in multiple contexts.

In addition, the same text string may retrieve words in multiple languages, adding yet more variance to the results. Better retrieval techniques will narrow the gap between searchers and creators of content, but will never close that gap completely.

Searching for information in multi-media digital libraries is more complex than text-only searching. Consider the many options for describing sounds, images, numeric data sets, and mixed-media objects. We might describe sounds with words, or with other sounds (e.g., playing a tune and finding one like it); we might describe an image with words, by drawing a similar object, or by providing or selecting an exemplar. As Croft (1995) notes in an earlier D-Lib issue, general solutions to multi-media indexing are very difficult, and those that do exist tend to be of limited utility.

The most progress is being made in well-defined applications in a single medium, such as searching for music or for photographs of faces. Cultural issues pervade digital library applications, whether viewing culture at the application level, such as variations in approaches to image retrieval by the art, museum, library, scientific, and public school communities, or on a multi-national scale, such as the differing policies on information access between the United States and Central and Eastern Europe. Designing digital libraries for distributed environments involves complex tradeoffs between tailoring to local cultures and meeting the standards and practices necessary for interoperability with other systems and services.

From Local Systems to Global Systems

The easiest systems to design are those for well-defined applications and well-defined user populations. Under these conditions, designers can build closed systems tailored to a community of users, iteratively testing and refining capabilities. These are rare conditions today, however. More often, we are designing open systems that serve not only a local population, but also remote and perhaps unknown populations. Examples include digital libraries of scholarly materials built by and for one university, then later made openly available on the Internet;

business assets databases developed and tested at a local site and then provided to corporate sites around the world; scientific databases designed for research applications, later made available for educational purposes; and library catalogs designed for a local university, later incorporated into national and international databases for resource sharing. Any of these applications could involve content in multiple media and multiple languages.

Design Tradeoffs

Consider how the design issues change from local to distributed systems. In local systems, designers can tailor user interfaces, representation of content, and functional capabilities to the local culture and to the available hardware and software. Input and output parameters are easily specified. If users need to create sounds or to draw, these capabilities can be provided, along with display, capture, and printing abilities in the matching standards. Keyboards can be set to support the local language(s) of input; screens and printers can be set to support the proper display of the local languages as well.

Designers have far less control over digital libraries destined for use in globally distributed environments. Users' hardware and software platforms are typically diverse and rapidly changing. Designers often must specify a minimum configuration or require a minimum version of client software, making tradeoffs between lowering the requirements to reach a larger population or raising requirements to provide more sophisticated capabilities. The more sophisticated the multi-media or multi-lingual searching capabilities, the higher the requirements are likely to be, and the fewer people that are likely to be served. While good design includes employing applicable standards, determining which standards are appropriate in the rapidly evolving global information infrastructure involves tradeoffs as well.

The use of some standards may be legislated by the parent organization or funding agency, and the use of other standards may be a matter of judging which are most stable and which are most likely to be employed in other applications with which the current system needs to exchange data. In the case of character sets for representing text in digital libraries, designers sometimes face a choice between a standard employed within their country to represent their national language and a universal character set in which their national language is more commonly represented in other countries.

At present, massive amounts of textual data are being generated in digital form, and represented in formats specific to applications, language, and countries. The sooner the digital library community confronts this tidal wave of "legacy data" in incompatible representations, the more easily this interoperability problem may be solved.

Representation in Digital Form

Although we have been capturing text, images, and sounds in machine-readable forms for several decades, issues of representation became urgent only when we began to access, maintain, exchange, and preserve data in digital form. In information technologies such as film, phonograph, CD-ROM, and printing, electronic data often is an intermediate format. When the final product was published or produced, the electronic data often were destroyed, and the medium (disks, tapes, etc.) reused. In digital libraries, the perspective changes in two important ways: (1) from static output to dynamic data exchange; and (2) from a transfer mechanism to a permanent archival form. In sound or print recordings, for example, once the record is issued or the book printed, it no longer matters how the content was represented in machine-readable form.

In a digital library, the representation matters because the content must be continuously searched, processed, and displayed, and often must be exchanged with other applications on the same and other computers. When electronic media were viewed only as transfer mechanisms, we made little attempt to preserve the content. Many print publications exist only in paper form, the typesetting tapes used to generate them long since overwritten. Much of the early years of television broadcasts were lost, as the recording media were reused or allowed to decay. Now we recognize that digital data must be viewed as a permanent form of representation, requiring means to store content in complete and authoritative forms, and to migrate content to new technologies as they appear.

Language and Character Sets

Character set representation is a problem similar to that of representing multi-media objects in digital libraries, yet is more significant due to the massive volume of textual communication and data exchange that takes place on computer networks. Culture plays a role here as well: speakers of all languages wish to preserve their language in its complete and authoritative form. Incomplete or incorrect data exchange results in failures to find information, in failures to authenticate identities or content, and in the permanent loss of information.

Handling character sets for multiple languages is a pervasive problem in automation, and one of great concern to libraries, network developers, government agencies, banks, multinational companies, and others exchanging information over computer networks. Much to the dismay of the rest of the world, computer keyboards were initially designed for the character set of the English language, containing only 26 letters, 10 numbers, and a few special symbols. While variations on the typical English-language keyboard are used to create words in most other languages, doing so often results in either (1) a loss of data, or (2) encoding characters in a language-specific or application-specific format that is not readily transferable to other systems.

Transliteration and Other Forms of Data Loss

Languages written in non-Roman scripts, such as Japanese, Arabic, Chinese, Korean, Persian (Farsi), Hebrew, and Yiddish (the "JACKPHY" languages), and Russian, are transliterated into Roman characters in many applications. Transliteration matches characters or sounds from one language into another; it does not translate meaning. Considerable data loss occurs in transliteration. The process may be irreversible, as variations occur due to multiple transliteration systems for a given language (e.g., Peking vs. Beijing, Mao Tse-tung vs. Mao Zedong (Chinese), Tchaikovsky vs. Chaikovskii (Russian)), and the transliterated forms may be unfamiliar to speakers of that language.

Languages written in extensions of the Roman character set, such as French, Spanish, German, Hungarian, Czech, and Polish, are maintained in incomplete form in some applications by omitting diacritics (accents, umlauts, and other language-specific marks) that distinguish their additional characters. These forms of data loss are similar to those of "lossy" compression of images, in which data are discarded to save storage costs and transmission time while maintaining an acceptable reproduction of the image. Any kind of data loss creates problems in digital libraries.

Variant forms of words will not match and sort properly, incomplete words will not exchange properly with digital libraries using complete forms, and incomplete forms may not be adequate for authoritative or archival purposes. The amount of acceptable loss varies by application. Far more data loss is acceptable in applications such as email, where rapid communication is valued over authoritative form, than in financial or legal records, where authentication is essential, for example.

Character Encoding

The creation of characters in electronic form involves hardware and software to support input, storing, processing, sorting, displaying, and printing. The internal representation of each character determines how it is treated by the hardware (keyboard, printer, VDT, etc.) and the application software. Two characters may appear the same on a screen but be represented differently due to their different sorting positions in multiple languages, for example. Conversely, the same key sequence on two different keyboards may produce two different characters, depending upon the internal representation that is generated.

Character encoding for digital libraries includes all of these aspects:

- The keyboard commands used to generate characters, especially characters with diacritics, for building the digital library content;
- The keyboard commands used to generate characters to search the digital library;
- Rules for sorting characters in correct alphabetic sequence, which are dependent on the internal representation of the character; the correct sequence varies by language;
- Display of characters on computer screens; and
- Output of characters on printers and other devices.

Numerous possibilities exist for mismatches and errors in access to digital libraries in distributed environments, considering the vast array of hardware and software employed by DLs and their users and the variety of languages and character encoding systems that may be involved.

Mono-lingual, Multi-lingual, and Universal Character Sets

Many standards and practices exist for encoding characters. Some are language-specific, others are script-specific (e.g., Latin or Roman, Arabic, Cyrillic), and "universal" standards that support most of the world's written languages are now available. Exchanging data among digital libraries that employ different character encoding formats is the crux of the problem. If mono-lingual DLs all use the same encoding format, such as ASCII for English, data exchange should be straightforward. If mono-lingual DLs use different formats, such as the three encoding formats approved for the Hungarian language by the Hungarian standards office, then data exchange encounters problems. Characters generated by a keyboard that is set for one encoding system may not match characters stored under another encoding system; characters with diacritics may display or print incorrectly or not display at all.

DLs using the same script-specific formats, such as Latin-2 extended ASCII that encompasses the major European languages, should be able to exchange data with each other. When DLs using Latin-2 attempt to exchange data in those same languages with DLs using language-specific formats, mismatches may occur. Similarly, mismatches may occur when DLs that employ Latin-2 for European languages exchange data with DLs that employ a different multi-lingual set such as the American Library Association character set commonly used in the United States. After many years of international discussion on the topic, Unicode appears to be emerging as the preferred standard to support most of the world's written languages.

A universal character set offers great promise for solving the data exchange problem. If data in all written languages are encoded in the same format, then data can be exchanged between mono-lingual and multi-lingual digital libraries. Just as the networked world is moving toward hardware platform-independent solutions, adopting Unicode widely would move us toward language-independent solutions to distributed digital libraries and to universal data exchange. Techniques for automatic language translation would be assisted by a common character set standard as well. Any solution that appears too simple probably is. Major hardware and software vendors are beginning to support Unicode, but it is not embedded in much application software yet. Unicode requires 16 bits to store each character — twice as much as ASCII, at 8 bits.

However, Unicode requires only half as much space as the earlier version of ISO 10646 (32 bits), the competing and more comprehensive universal character set. Unicode emerged as the winner in a long standards battle, eventually merging with ISO 10646, because it was seen as easier to implement and thus more likely to be adopted widely. As storage costs continue to decline, the storage requirements of Unicode will be less of an issue for new applications. Massive amounts of text continue to be generated not only in language-specific and script-specific encoding standards, but in local and proprietary formats. Any of this text maintained in digital libraries may become "legacy data" that has to be mapped to Unicode or some other standard in the future.

At present, digital library designers face difficult tradeoffs between the character set standards in use by current exchange partners, and the standard likely to be in international use in the future for a broader variety of applications.

Library Community Approaches

The international library community began developing large, multi-language digital libraries in the 1960s. Standards for record structure and character sets were established long before the Internet was created, much less Unicode. Hundreds of millions of bibliographic records exist around the world in variations of the MARC (MAchine Readable Cataloging) standard, although in multiple character set encoding formats. OCLC Online Computer Library Center, the world's largest cataloging cooperative, serves more than 17,000 libraries in 52 countries and contains over 30 million bibliographic records with over 500 million records of ownership attached in more than 370 languages. OCLC uses the American Library Association (ALA) character set standard, which extends the English-language keyboard to include diacritics from major languages. Text in most other languages is maintained in transliterated form.

The Library of Congress, which contributes its records in digital form to OCLC, RLIN (Research Libraries Information Network, the other major U.S.-based bibliographic utility), and other cooperatives, also do original script cataloging for the JACKPHY languages mentioned earlier. RLIN pioneered the ability to encode the JACKPHY languages in their original script form for bibliographic records, using available script-specific standards. Records encoded in full script form are exchanged between the Library of Congress, RLIN, OCLC, other bibliographic utilities in the U.S. and elsewhere, and many digital libraries maintained by research libraries. Catalog cards are printed in script and Romanized forms from these databases, but direct use of the records in script form requires special equipment to create and display characters properly. Records from OCLC, RLIN, and other sources are loaded into the online catalogs of individual libraries, where they usually are searchable only in transliterated forms.

Some online catalogs support searching with diacritics, while others support only the ASCII characters. Regardless of the local input and output capabilities, if characters are represented internally in their fullest form, they will be available for more sophisticated uses in the future when the search and display technologies become more widely available. Libraries always have taken a long-term perspective on preserving and providing access to information. They manage content in many languages and cooperate as an international community to exchange data in digital form.

Thus it is not surprising that libraries were among the first institutions to tackle the multi-lingual character set problem. Over the last 30 years, libraries have created massive stores of digital data. Not only do libraries create and maintain new bibliographic records in digital form, a growing number of the world's major research libraries have converted all of their historical records — sometimes dating back several hundred years — into a common record structure. By now, libraries have the expertise and influence to affect future developments in standards for character sets and other factors in data exchange.

The library world is changing, however, as new regions of the world come online. The European Union is promoting Unicode and funding projects to support Unicode implementation in library automation. Automation in Central and Eastern Europe (CEE) has advanced quickly since 1990. A survey of research libraries in six CEE countries, each with its own national language and character set, indicates that a variety of coding systems are in use. As of late 1994, more than half used ASCII Latin2, one used Unicode, and the rest used a national or system-specific format; none used the ALA character set. The national libraries in these countries are responsible for preserving the cultural heritage of their countries that appears in published form, and thus require that their language be preserved in its most complete and authoritative digital form.

Transliterated text or characters stripped of diacritics are not acceptable. Several of these national libraries are now working closely with OCLC, toward the goal of exchanging data in authoritative forms. As libraries, archives, museums, and other cultural institutions throughout the world become more aware